TECHNICAL REPORT

Methodologies in Analyzing the Root Causes of Nunn-McCurdy Breaches

Irv Blickstein • Jeffrey A. Drezner • Brian McInnis • Megan McKernan
Charles Nemfakos • Jerry M. Sollinger • Carolyn Wong

Prepared for the Office of the Secretary of Defense

Approved for public release; distribution unlimited

RAND NATIONAL DEFENSE RESEARCH INSTITUTE

The research described in this report was prepared for the Office of the Secretary of Defense (OSD). The research was conducted within the RAND National Defense Research Institute, a federally funded research and development center sponsored by OSD, the Joint Staff, the Unified Combatant Commands, the Navy, the Marine Corps, the defense agencies, and the defense Intelligence Community under Contract W74V8H-06-C-0002.

Library of Congress Cataloging-in-Publication Data is available for this publication.

ISBN: 978-0-8330-7630-4

Published 2012 by the RAND Corporation
1776 Main Street, P.O. Box 2138, Santa Monica, CA 90407-2138
1200 South Hayes Street, Arlington, VA 22202-5050
4570 Fifth Avenue, Suite 600, Pittsburgh, PA 15213-2665
RAND URL: http://www.rand.org/
To order RAND documents or to obtain additional information, contact
Distribution Services: Telephone: (310) 451-7002;
Fax: (310) 451-6915; Email: order@rand.org

Preface

As a result of continuing concern with large cost overruns in a broad range of major defense programs, Congress enacted new statutory provisions extending the ambit of the existing Nunn-McCurdy Act. In accordance with the revised Nunn-McCurdy law, the Performance Assessments and Root Cause Analysis (PARCA) office must provide its root cause explanation as part of a 60-day program review triggered when the breach is reported by the applicable military department secretary.

In March 2010, the newly created PARCA office within the Office of the Secretary of Defense (OSD), in view of staffing limitations, elected to rely on federally funded research and development center (FFRDC) support to help discharge its new responsibilities. It engaged the RAND Corporation to conduct multiple studies on the root causes of Nunn-McCurdy breaches or other large cost increases in six major defense acquisition programs: the Wideband Global Satellite, the Longbow Apache, the DDG-1000, the Joint Strike Fighter, the Excalibur, and the NAVY Enterprise Resource Planning (ERP). The results appear in two reports.[1]

In the course of conducting the six root cause analyses, RAND researchers gained considerable insight into the methodology of such analyses and the data sources necessary to support them, and this technical report describes their findings.

This research was sponsored by OSD PARCA and conducted within the Acquisition and Technology Policy Center of the RAND National Defense Research Institute, a federally funded research and development center sponsored by the Office of the Secretary of Defense, the Joint Staff, the Unified Combatant Commands, the Navy, the Marine Corps, the defense agencies, and the defense Intelligence Community.

For more information on the RAND Acquisition and Technology Policy Center, see http://www.rand.org/nsrd/ndri/centers/atp.html or contact the director (contact information is provided on the web page).

---

[1] Irv Blickstein et al., *Root Cause Analyses of Nunn-McCurdy Breaches,* Volume 1: Zumwalt-*Class Destroyer, Joint Strike Fighter, Longbow Apache, and Wideband Global Satellite,* Santa Monica, Calif.: RAND Corporation, MG-1171/1-OSD, 2011; Irv Blickstein et al., *Root Cause Analyses of Nunn-McCurdy Breaches,* Volume 2: *Excalibur Artillery Projectile and the Navy Enterprise Resource Planning Program, with an Approach to Analyzing Complexity and Risk,* Santa Monica, Calif.: RAND Corporation, MG-1171/2-OSD, 2012.

Contents

Figures

Tables

Summary

Background

Continuing program cost growth and observations by the Government Accountability Office (GAO) placing defense acquisition on the high-risk target list raised concern in Congress about the execution of major defense acquisition programs. This concern and the reality of shrinking defense budgets led Congress to enact statutory provisions that would increase the focus of senior policymakers on oversight of major defense acquisition programs (MDAPs) and other large costly programs.[2] The Weapon Systems Acquisition Reform Act (WSARA) of 2009[3] established a number of requirements that affected the operation of the defense acquisition system and the duties of the key officials who support it, including the requirement to establish a new organization in the Office of the Secretary of Defense (OSD) with the mandate to conduct and oversee performance assessments and root cause analyses (PARCA) for MDAPs.

In March 2010, the director of PARCA determined that he needed support to execute his statutory responsibilities and turned to federally funded research and development centers (FFRDCs) and academia to provide that support for the research and analysis of program execution status. RAND was one FFRDC engaged to perform research and analysis and provide recommendations and was originally assigned responsibility for four programs.[4] After completing that initial effort, RAND was assigned two additional programs for research and analysis: Excalibur and Navy Enterprise Resource Planning (ERP).

Purpose

This technical report documents the methodology RAND developed to carry out the root cause analyses (RCAs). In analyzing six programs, RAND has developed some expertise in what is required to carry out these analyses effectively. It is important

---

[2] Ike Skelton Defense Authorization Act for Fiscal Year 2011, December 20, 2010.

[3] Public Law 111-23, May 22, 2009, Weapon Systems Acquisition Reform Act (WSARA) of 2009.

[4] Blickstein et al., 2011.

to chronicle the approach used for all the analyses so that others may use it in their own analytic efforts. The report also gathers together extensive documentation on data sources that can be used for root cause analyses and for other purposes pertaining to the six specific programs RAND analyzed.

Observations on the Conduct of Root Cause Analyses

Each acquisition program is unique, and each RCA is unique. However, RAND's experience in conducting six root cause analyses indicates that a set of core activities is instrumental to a successful effort. These activities define a generic root cause methodology whose key components include the following:

- Gather and review readily available data.
- Develop a hypothesis.
- Set up long-lead-time activities.
- Document the unit cost threshold breach.
- Construct a time line of relevant cost growth events in the program history.
- Verify the cost data and quantify cost growth.
- Create and analyze the program cost profiles pinpointing occurrences of cost growth.
- Match the time line events with changes in the cost profiles and derive root causes of cost growth.
- Reconcile any remaining issues.
- Attribute unit cost growth to root causes.

Successful execution of this set of activities should enable the research team to create the primary deliverables and postulates for a root cause analysis: a summary narrative that includes clearly stated root causes of cost growth supported by a formal documentation of the cost threshold breach, a summary time line of program events leading to the Nunn-McCurdy breach, funding profiles, a completed PARCA office–generated root cause matrix, and a breakdown of the amount of cost growth attributable to each root cause; a briefing that corresponds to the narrative; and a full root cause report.

In addition to developing deliverables and postulates, the RCA process is designed to improve the research focus iteratively. At each stage of the RCA, information is both drawn from and contributed to the program archive. The RCA analytic team can use this insight not only to improve the interim products that result from successive stages of the RCA but also to advance the original hypothesis that guides the research. This process of regularly refining the guiding hypothesis with the insights gained during the production of key deliverables and postulates enables the research team to identify quickly the root causes of a program's failure.

Acknowledgments

We would like to acknowledge the helpful interactions we had with Director Gary Bliss and David Nicholls of the PARCA office.

The RAND team also enjoyed, and benefited from, informal collaboration with colleagues at the Institute for Defense Analyses. RAND colleagues familiar with the programs we studied also provided us with helpful support.

We express our appreciation for the good work done by our two reviewers, Jack Graser and Mike Hammes. Their thorough and thoughtful reviews made this a better report.

Finally, we thank Jennifer Miller and Alex Chinh, who provided administrative support and technical support for the project.

Abbreviations

AB3	Apache Longbow Helicopter, Block III
ACAT	Acquisition Category
ADM	Acquisition Decision Memorandum
AIS	Atlantic Inertial Systems
APB	acquisition program baseline
APUC	average procurement unit cost
AT&L	Acquisition Technology and Logistics
BES	Budget Estimate Submission
BY	base year
CAIG	Cost Analysis Improvement Group
CAPE	Cost Assessment and Program Evaluation
CARD	Cost Analysis Requirements Description
CBO	Congressional Budget Office
CCA	component cost analysis
CPARS	Contractor Performance Assessment Reporting System
CRS	Congressional Research Service
DAB	Defense Acquisition Board
DAE	Defense Acquisition Executive
DAES	Defense Acquisition Executive Summary
DAMIR	Defense Acquisition Management Information Retrieval
DCMA	Defense Contract Management Agency
DD&C	detailed design and construction

DDG	guided missile destroyer
DoD	Department of Defense
EDM	Engineering Design Model
EMD	Engineering Manufacturing and Development
ERP	Enterprise Resource Planning
EVM	earned value management
FFP	firm fixed price
FFRDC	federally funded research and development center
FY	fiscal year
GAO	Government Accountability Office
HTS	Human Terrain System
ICE	independent cost estimate
IOC	initial operational capability
IOT&E	initial operational test and evaluation
IPT	Integrated Product Team
JIC	Joint Inflation Calculator
JPO	Joint Program Office
JROC	Joint Requirements Oversight Council
JSF	Joint Strike Fighter
LCCE	life-cycle cost estimate
LRIP	low rate initial production
MAIS	Major Automated Information System
MAR	Major Automated Information System Annual Report
MDAP	major defense acquisition program
MOD	Ministry of Defence
MQR	Major Automated Information System Quarterly Report
MS 0	Milestone 0
MS B	Milestone B
N-M	Nunn-McCurdy
OIF	Operation Iraqi Freedom

OIPT	Overarching Integrated Process Team
OPEVAL	operational evaluation
ORD	Operational Requirements Document
O&S	operations and support
OSD	Office of the Secretary of Defense
OUSD	Office of the Under Secretary of Defense
PARCA	Performance Assessments and Root Cause Analyses
PAUC	program acquisition unit cost
PDR	Program Deviation Report
PEO	Program Executive Officer
PEO-AMMO	Program Executive Officer for Ammunition
POM	Program Objective Memorandum
RCA	root cause analysis
RDT&E	Research Development Test and Evaluation
SAR	Selected Acquisition Report
SDD	System Development and Demonstration
SECDEF	Secretary of Defense
SECNAV	Secretary of the Navy
SSSFM	Spin Stabilized Sensor Fused Munition
SYSCOMS	System Commands
USAF	U.S. Air Force
USC	U.S. Code
USD	Under Secretary of Defense
VCSA	Vice Chief of Staff of the Army
WCF	Working Capital Fund
WGS	Wideband Global Satellite
WSARA	Weapon Systems Acquisition Reform Act

CHAPTER ONE

Introduction

Background

The Congress of the United States has long been concerned about the cost of acquiring weapon systems for the military services. Major defense acquisition programs (MDAPs) are sometimes plagued by cost overruns. Over the last quarter century, Congress has imposed both structure and process on defense programs with an eye to fostering better management of them. Those measures have not delivered the results Congress anticipated. Therefore, Congress has once again attempted to improve the acquisition of weapons by imposing statutory requirements on the Department of Defense (DoD), this time including assessments of acquisition elements in an organization that employs members of the acquisition workforce, carries out acquisition function, and focuses primarily on acquisition.[1] The Weapon Systems Acquisition Reform Act (WSARA) of 2009[2] established a number of requirements that affected the operation of the defense acquisition system and the duties of the key officials who support it, including the requirement to establish a new organization in the Office of the Secretary of Defense (OSD) with the mandate to conduct and oversee performance assessments and root cause analysis (PARCA) for MDAPs.

Congress enacted new statutory provisions extending the reach of the existing Nunn-McCurdy Act. In accordance with the revised Nunn-McCurdy law, the PARCA office must provide its root cause explanation as part of a 60-day program review triggered when a military department secretary reports a Nunn-McCurdy breach.

Purpose

This technical report provides a detailed discussion of how to proceed with a root cause analysis and, equally important, a listing of the data sources needed to underpin such an analysis. It also provides a detailed listing of the data sources used to support

---

[1] Ike Skelton Defense Authorization Act for Fiscal Year 2011, HR 6523, Section 861.

[2] Public Law 111-23, Weapon Systems Acquisition Reform Act (WSARA) of 2009, May 22, 2009.

our analyses of the six programs RAND has analyzed thus far: the *Zumwalt*-class Destroyer, Joint Strike Fighter, Longbow Apache, Wideband Global Satellite (WGS), Excalibur, and the Navy Enterprise Resource Program. These sources will prove invaluable to anyone wishing to study the procurement history of these programs.

Organization of This Report

The report is organized into four chapters and seven appendixes. Chapter Two describes the root cause analysis (RCA) methodology that RAND researchers developed over the course of conducting the six RCAs requested by the Department of Defense. It describes both the process and the products. Chapter Three catalogs the sources of information that can inform the RCA. Chapter Four presents our conclusions.

The seven appendixes list the documents that pertain to each RCA completed by RAND thus far.

We note that Appendix C arrays data and data sources relating to the *Zumwalt*-class Destroyer DDG-1000 root cause analysis in a way that enables the systematic planning activity needed to use the bibliography.

Root Cause Analysis Methodology

There is no standard step-by-step method for conducting an RCA—each is unique because each program is unique. However, key elements of the program's background should be analyzed in the course of conducting an RCA. These key elements define a generic methodology for conducting a successful root cause analysis. Figure 2.1 illustrates the path a typical RCA team would navigate to identify the root causes of a program's cost growth. Although this illustration and the discussion in this chapter are sequential, the process is iterative. In addition, many of the activities described below usually occur simultaneously because only a short time period is allowed by law for

Figure 2.1
Generic RCA Methodology

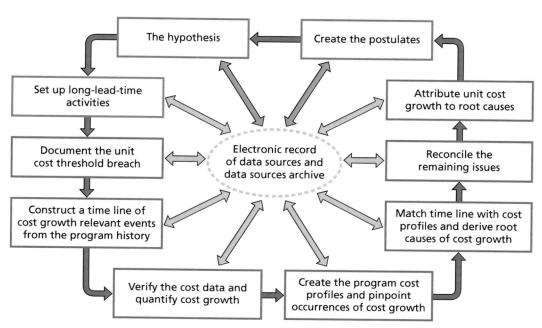

NOTE: The green arrows indicate the start and stop points of the cycle.
RAND *TR1248-2.1*

completion of a root cause analysis, subsequent decision (recertification or termination of the program) by the Secretary of Defense, and formal notification to Congress of the Secretary of Defense's decision.[1]

Figure 2.1 shows that activities conducted during a root cause analysis should include the following:

- Analyze the letter of breach notification and develop a hypothesis.
- Set up long-lead-time activities.
- Document the unit cost threshold breach.
- Construct a time line of relevant cost growth events in the program history.
- Verify the cost data and quantify cost growth.
- Create and analyze the program cost profiles pinpointing occurrences of cost growth.
- Match the time line events with the changes in the cost profiles and derive root causes of cost growth.
- Reconcile remaining issues.
- Attribute unit cost growth to root causes.
- Create the postulates, deliverables, and reformation of the original hypothesis for the PARCA office.
- Revisit the hypothesis to determine its validity.

The recording of bibliographic information for the data sources archive and creation of the data sources database are ongoing activities throughout the RCA. Detailed explanations of each activity in the RCA methodology follow the synopsis.

In the RCAs performed to date, the PARCA office has requested several deliverables:

- a completed root cause matrix in the format supplied by the PARCA office
- a summary narrative
- a set of briefing charts based on the narrative
- a full RCA report.

All deliverables except the full RCA report should be supplied by PARCA office deadlines to ensure that these materials can be used to support the recertification decision.

To ensure uniformity in how root causes are reported, the PARCA office has provided a root cause matrix to summarize the root causes of Nunn-McCurdy breaches.

---

[1]   10 U.S. Code (USC) § 2433(a) states that notification to Congress of program recertification by the Secretary of Defense is required before the end of the 60-day period that begins on the day the next Selected Acquisition Report is required by 10 USC § 2432(f). See Chapter Three for further discussion of the RCA environment.

A set of briefing charts based on the narrative is the third product delivered before the recertification decision. One slide should show the completed PARCA root cause matrix.

The final full RCA report can be delivered after the recertification due date. This report should contain the details of each activity conducted during the program RCA. A copy of the data archive and access to the electronic version as well as a copy of the data sources database are typically delivered with the final full RCA report.

The remainder of this chapter discusses each element of the methodology in greater detail.

The Hypothesis

When a program incurs a Nunn-McCurdy breach, the military department formally notifies Congress of this breach and, as part of the notification, states the reason for it. The reason stated is the initial hypothesis that analysts use to begin an RCA. At the conclusion of the RCA, the hypothesis is revisited to determine its validity.

Set Up Long-Lead-Time Activities

Program office personnel, including government officials (former and current) and contractors, provide valuable first-hand accounts of a program's successes, failures, strengths, and weaknesses. Insight provided by program office personnel is not always captured in official documentation. Past RCA teams have sought to interview program managers, contractors, user groups, government financial personnel, and other government executives (e.g., Comptroller of the Navy). Interviews with government executives, program office personnel, and contractors, particularly current and former program managers and their deputies, provide unique perspectives on the programs. Given the busy schedules of program office personnel, setting up such meetings can require considerable lead time and sometimes coordination with PARCA office personnel. Locating, contacting, and securing agreements to interview former program managers and knowledgeable former government executives can also require many days. Hence, previous RAND RCA teams pursued these interviews at the beginning of an RCA. They then used the lead time to gain an understanding of the program issues through the activities described below. Post-interview discussions were then held to help resolve any issues remaining in the latter stages of the RCA after initial interviews took place.

Although no particular technique will guarantee interviews with all persons of interest, past RCA teams have typically secured interviews with candidates and then, at the end of each interview, requested that the interviewee assist the RCA team with securing interviews with other pertinent personnel. Subsequent requests for interviews were then made using past interviewees as references. This technique can quickly open doors that might otherwise require more lengthy processes.

Document the Unit Cost Threshold Breach

One of the first analytic steps in an RCA is to develop an understanding of the Nunn-McCurdy breach that triggered the RCA. To do so, the RCA team must document the "speeding ticket" from official documents and data. The speeding ticket should specify whether the average procurement unit cost (APUC) or program acquisition unit cost (PAUC) exceeded critical thresholds and whether the original baseline or the current baseline was breached. In addition, the cost growth expressed as percentage and dollar amounts over original or current baseline should be calculated. Brief explanations of the immediate causes of the unit cost growth can also be summarized. Table 2.1 (below) shows a suggested format for documenting the unit cost growth breach. The format of the matrix issued by the PARCA office is based on Public Law 111–23, Title I, § 103. Details on creating the speeding ticket are given below.

A Program Deviation Report will generally provide one of the first official indications that a program is likely to incur a critical unit cost growth breach. This report will generally provide a brief description of the immediate unanticipated change in program execution that led to the report. The report will likely reference a previous official notification such as an Acquisition Decision Memorandum (ADM) as the source for why an unanticipated change in the program execution occurred. The percentage and dollar amounts of the unit cost growth would typically be included in a Program Deviation Report.

The military department secretary of the program's service officially notifies Congress of the unit cost growth breach in letters to the chairman and ranking members of the following committees:

- Senate Committee on Appropriations
- House Committee on Appropriations
- House Committee on Armed Services
- Senate Committee on Armed Services.

These letters generally describe the breach, provide some explanation of what led to it,[2] and are useful in developing an initial working hypothesis that helps govern the approach to analysis. The descriptions of unit cost growth expressed in the letters

---

[2] These letters of notification to Congress usually contain identical information. Hence, examining one such letter is sufficient—no new information would typically be available in the other notification letters. See Chapter Three for more details on these letters and the Program Deviation Reports.

should be compared to the analogous figures in the Program Deviation Reports. Any mismatches should be noted as issues that need to be resolved before the speeding ticket is finalized.

The latest available Selected Acquisition Report (SAR) or Major Automated Information System (MAIS) Annual Report (MAR) or MAIS Quarterly Report (MQR) on the program will contain cost data on unit costs that can be used to compute the dollar amounts and percentages that should be shown on the speeding ticket. If the latest available SAR was issued after the Nunn-McCurdy breach, the SAR will show the breach along with the pertinent dollar and percentage increases. If the latest available SAR was issued before the Nunn-McCurdy breach (e.g., December of the year before the breach), then the latest Defense Acquisition Executive Summary (DAES) may be the first official cost report that reflects the breach. Since DAESs are issued monthly, a DAES that reflects the Nunn-McCurdy breach may be available before a SAR (issued quarterly or annually) that reflects the breach. More details on the SAR, MAR, and MAIS are given in Chapter One.

Table 2.1 suggests a format for the speeding ticket. If the program Nunn-McCurdy breach was only for the APUC, then only the second and fourth rows of the table need be included. Likewise, if the program Nunn-McCurdy breach was to the PAUC, only the first and third rows of the table need be included. If the program incurred critical breaches to both the APUC and PAUC, then the information in all of the rows in Table 2.1 should be included. Different formats can be used for the speeding ticket, but all of the information shown in the suggested format in Table 2.1 should always be included. For this reason, we explain each component of the speeding ticket in the following paragraphs.

The first column of Table 2.1 shows the program name.

The second column shows the baseline unit costs as shown in the current Acquisition Program Baseline (APB) and the original APB. These figures are included in the SAR, MAR, and DAES. The current APB figures and the original APB figures may be expressed in different fiscal year (FY) dollars. In such a case, the analyst should convert the original APB fiscal year dollars to current APB year dollars, so that all amounts on the speeding ticket will be comparable.[3] The Joint Inflation Calculator (JIC)[4] can be used to convert dollar amounts.

The third column shows the current unit cost estimates, which should reflect the Nunn-McCurdy breach. They can normally be found on the latest SAR or DAES that reflects the critical Nunn-McCurdy breach. The source for these unit costs should be

[3] In some cases, there may be reasons to choose different fiscal year dollars. As long as all amounts on the speeding ticket are expressed in the same fiscal year dollars, the amounts will be comparable.

[4] The initial version of JIC was prepared by the Naval Center for Cost Analysis to provide Army and Department of the Navy inflation rates and indexes for the cost-estimating community. The JIC exists in the form of a Microsoft Excel Workbook. The latest version is available for download; see Naval Center for Cost Analysis, "NCCA Inflation Indices," January 2010.

Table 2.1
Format for the Speeding Ticket

Program	Baseline Unit Cost (FY $ millions)	Current Estimate (Source, e.g., Dec09 SAR) FY 20XX $ millions	Baseline Breached	Cost Growth Threshold Breaches						
				Percentage	Amount	Level	Baseline Quantity	Current Quantity	Cause in Source	Explanation in Source
Program name	APUC $XXX (year of current APB)	APUC $XXX	Over current baseline (year of current APB)	APUC +XXX%	+$XXX FY XXX $M	Critical	#	#		
	PAUC $XXX (year of current APB)	PAUC $XXX		PAUC +XXX%	+$XXX FY XXX $M	Critical	#	#		
	APUC $XXX (year of original APB)	APUC $XXX	Over original baseline (year of original APB)	APUC +XXX%	+$XXX FY XXX $M	Critical	#	#		
	PAUC $XXX (year of original APB)	PAUC $XXX		PAUC +XXX%	+$XXX FY XXX $M	Critical	#	#		

SOURCES: List sources used to complete table.

NOTE: The numbers in red indicate the "speeding ticket" triggering root cause analysis by PARCA.

entered in the third column of the header as shown in Table 2.1. Again, the JIC can be used to convert amounts to current APB fiscal year dollars.

The fourth column shows which baseline (current APB or original APB) was critically breached.

The fifth column shows the percentage of the critical APUC or PAUC breach(es), which can be found in the latest SAR, MAR, or DAES that reflects the critical Nunn-McCurdy breach(es), or it can be calculated using the amounts in the baseline unit cost and current estimate columns.

The sixth column shows the cost growth in dollars. These amounts can be taken from the SAR or DAES or computed using the amounts in the baseline unit cost and current estimate columns.

The seventh column records that the breach is a critical Nunn-McCurdy breach. If the analyst chooses to display these on the speeding ticket in addition to the critical breaches, the distinction between the two types of breaches can be shown in this column.

The eighth column shows the baseline quantity, which is found in the SAR, MAR, and DAES.

The ninth column shows the current quantity after the Nunn-McCurdy breach. This amount can be found in the SAR, MAR, or DAES that reflects the Nunn-McCurdy breach. Sometimes, particularly when the Nunn-McCurdy breach occurs because a quantity has changed, official documents such as the Program Deviation Report, the ADM documenting the critical Nunn-McCurdy breach, or official letters to Congress will show the current quantity.

The tenth column can summarize any cause expressed in the SAR, MAR, DAES, Program Deviation Report, ADM, or letters to Congress. For example, the SAR may state that a revised program cost estimate led to the unit cost breach.

The eleventh column can summarize any explanation for the critical unit cost growth found in official documentation. For example, the SAR may state that a revised program cost estimate was necessitated by an unanticipated technical problem.

Sources used to create the speeding ticket can be listed in a source element at the bottom of the ticket.

Construct a Time Line of Relevant Cost Growth Events from the Program History

The speeding ticket provides a complete documentation of the charge the program is expected to respond to with the root cause analysis. Once the charge is understood, the team needs to construct a time line of key program events. Key events in the program's history are activities or occurrences that mark progress, problems, or issues leading up to the Nunn-McCurdy breach. Examples include milestones, ADMs, and changes in cost, quantity, or schedule. Included events should be accompanied with short notes on any qualifying factors marking those events. For example, "production halted for three months" might be qualified with a note that this event was due to a prolonged strike. Events relevant to the Nunn-McCurdy breach should be noted in the time line.

Events will be taken primarily from the program history, but some external events can also be pertinent. For example, changes in the industrial base can be caused by external events that may affect a program's acquisition strategy. Since events in and affecting a program are intricately linked, identifying the string of events that finally resulted in the critical unit cost breach is imperative to understanding the root causes of the critical cost growth. In some cases such as Excalibur, the string of events was embedded in the program history itself, and the RCA team identified and documented the string of events leading to the Nunn-McCurdy breach by requesting and examining detailed budget material as well as Army studies conducted by the Army's Center for Army Analysis. In the Excalibur case, the string of relevant events showed that the requirements changed after initiation of the program. In the Navy ERP case, the string of events showed that the business processes changed after the initiation of the program. In the case of the WGS,[5] the evolution of the commercial satellite industry was a key driver of unit cost growth in that program. In that case, the RCA team meshed the evolution of the commercial satellite industry with the WGS program history to show that the evolution led to the Nunn-McCurdy breach. Interviews and satellite industry literature were used as sources of information.

Construction of the relevant cost growth time line is likely to be an iterative task. The SAR (or MAR and MQR for MAIS programs) is a fruitful place to begin. All program SARs and MARs contain a summary of the program's history, and a preliminary time line of events can be constructed by going through all of a program's annual SARs. An outline of the program's history relevant to cost growth can also be created from these descriptions by distinguishing events that are pertinent to cost growth.

Although key events are usually included in the SAR and MAR histories, elaboration of events that may be pertinent to cost growth is often not given. To determine whether an event may have affected cost, the cost-reporting parts of the SAR can be reviewed. Notes are often present in the cost sections of the SAR that illuminate the reasons for cost changes.

Once a preliminary outline of relevant events is formed, other sources will need to be consulted to determine if and how the events affected cost. Sources to consult to update and revise the time line include Government Accounting Office (GAO) reports, Congressional Research Service (CRS) reports, Congressional testimony on the program, signed acquisition strategies, Program Deviation Reports, program briefings, service briefings on the program, and Nunn-McCurdy Overarching Integrated Process Team (OIPT) cost and management briefings. These sources may also iden-

---

5   Blickstein et al., 2011.

tify events pertinent to cost growth that are not mentioned in the SARs or MARs but should be added to the preliminary time line.

When the program time line has reached a stable state (sources do not reveal additional events or explanations to add to the time line), the resulting time line should read as a summary of the program's key events with those events that affected cost highlighted or distinguished in some manner. At this point, the analyst should review the time line and note those events that do not appear to be fully explained. For example, a quantity change with an incomplete or missing explanation of the reason for it should be flagged. At a minimum, the time line should show major milestones and all substantive quantity, cost, and schedule changes as well as the events that led to each change. The reason for including events that led to such changes is that quantity, schedule, and cost changes follow other events—these events do not occur without precursor events. Equally important in this time line review is the elimination of events that are not pertinent to cost growth. The detailed explanations will help the analyst decide if an event influenced cost. When in doubt, the event should remain in the time line until eliminated by further examinations, as described below.

Although there is no surefire way to determine what is important enough to include in the time line at the onset of an RCA, analysts can examine the program data to look for more obvious likely elements of interest such as the critical components that pose the greatest risk of program failure.

Verify the Cost Data and Quantify Cost Growth

The objective of this task is to make sure that the data in the official sources are consistent and reproducible. The RCA team should calculate the unit costs from the quantity and procurement cost figures to verify that the unit costs calculated from different sources are consistent. For example, unit costs calculated from the latest SAR should match unit costs calculated from the latest Program Objectives Memorandum (POM) data, and both should be consistent with any comparable data provided to the RCA team by the program office or by Cost Assessment and Program Evaluation (CAPE). Sources for discrepancies should be investigated and explanations noted. For example, if the quantities produced per year differ among sources, then the unit costs may also differ. Quantities may differ because the counts were made at different times. In such a case, the RCA team should investigate why the quantities are different. For example, one source might use the unit fielding date as date produced whereas another source might use the date on the Material Inspection and Receiving Report (Form DD250). If the dates differ in fiscal or calendar year, then both "counts" may be correct because they are calculated from different bases.

Create the Program Cost Profiles and Pinpoint Occurrences of Cost Growth

RDT&E Funding Profiles

When a program is initiated, program officials have a plan on how the program will progress toward developing and producing the final product. That plan is reflected in the Research, Development, Test, and Evaluation (RDT&E)[6] funding profile and the procurement funding profile. These profiles depict the program as envisioned. Each year, starting with the year the program enters Milestone B, the SAR updates the program profiles to reflect the President's Budget for the upcoming fiscal year. Plotting the RDT&E funding profiles from all years on a single graph visually depicts changes in how the program was originally planned and how that plan changed during the program's history. The RDT&E profiles graph enables the analyst to identify quickly where cost changes occurred during development and the magnitude of the cost changes. Obviously, cost increases are of primary concern, and the analyst needs to note the changes in the RDT&E profile and search for reasons for them. The SARs often include explanations for cost changes and also categorize RDT&E cost changes into one of seven categories: economic, quantity, schedule, engineering, estimating, support, and other. The RCA team can use these indicators to construct explanations for each cost change and note or calculate what percentage of the RDT&E cost growth is due to each cause. The SAR includes cost data in enough detail for the analyst to determine the percentage of cost change that can be attributed to each of the seven cost change categories.

Procurement Funding Profiles

In a manner analogous to the RDT&E funding profiles, the procurement funding profiles depict the procurement plan for the program. Changes in the procurement profiles over the course of the program's history indicate times and magnitudes of procurement issues. The analyst needs to note these changes and search for reasons for them. Again, the SARs often include explanations for cost changes and also categorize procurement cost changes into the same categories used with RDT&E: economic, quantity, schedule, engineering, estimating, other, and support. The RCA team can use these indicators to construct explanations for each cost change and note or calculate what percentage of the procurement cost growth is due to each cause.

Unit Cost Profiles

As the development and procurement funding profiles change, the APUC and PAUC can change as well. Each SAR contains a section that presents the current APUC and

---

[6]   RDT&E is often referred to as *development* in official documents such as the SAR. We use the terms interchangeably in this document as well.

PAUC. Plotting the APUC and PAUC profiles for all of the years since Milestone B will show how these cost elements have changed. It is often useful to superimpose the APUC and PAUC profiles on a bar graph of total program (RDT&E plus procurement) quantity for each year since Milestone B. Figure 2.2 shows such a graph for the Excalibur program.

Such a depiction will show how funding has changed and whether unit cost changes are consistent with program funding changes. Inconsistencies should be noted. At this point, the analyst may also consult budget material relating to the program. Budget documents often list all sources of funding including supplemental funding that may or may not be included in the SARs.[7] Substantial supplemental funding that is not included in SARs can indicate that the actual unit cost might be different from that reported in the SARs. Such instances should be noted.

Once again, the SARs often include explanations for cost changes and also categorize RDT&E and procurement cost changes into one of seven categories: economic, quantity, schedule, engineering, estimating, other, and support. The RCA team can use these indicators to construct explanations for each unit cost change and note or calculate what percentage of the procurement cost growth is due to each cause.

Figure 2.2
Example of Unit Cost and Quantity Graph for Excalibur

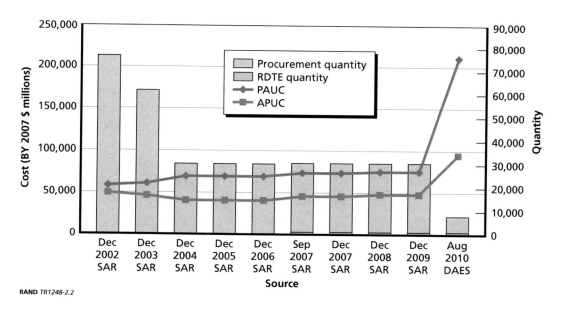

RAND *TR1248-2.2*

---

[7]   SARs may not be required to include funding from all sources.

Match Time Line with Cost Profiles and Derive Root Causes of Cost Growth

This portion of the methodology merges the time line with the cost profiles. The time line contains events that have been marked as likely to affect unit cost and events that are not yet fully explained. Events in the time line that have been noted as likely to affect cost should be matched with cost changes in the RDT&E profiles, procurement profiles, and unit cost profiles. In this matching exercise, the analyst is verifying that the summaries of the time line events adequately account for the cost changes in the funding profiles in terms of both timing and magnitude. For example, a change in the RDT&E funding profile might be traced to a schedule change in the time line that can itself be traced back to a technical issue. This sequence of events allows the analyst to explain an RDT&E funding profile change as being brought about by an earlier technical issue that required more time to be resolved and hence a subsequent schedule stretch and then a resulting development funding profile change. The analyst needs to make sure that explanations are in line with the magnitude of the cost changes indicated in the funding profiles. For example, if the time line shows that a requirements change led to a quantity change of about a 50 percent reduction, but the unit cost profiles show a unit cost change of only 5 percent in the years following the quantity reduction, then the analyst needs to seek further explanation for reconciling the two changes. Explanations may be embedded in other events such as the requirements change that led to the quantity reduction or changes in the industrial base. The RCA team may need to delve into technical issues or may need to consult supplementary material such as Defense Contract Management Agency (DCMA) reports to further trace and identify the sources of cost changes.

At the conclusion of this matching exercise, the analyst should have derived the probable root causes of cost growth. The analyst may be left with some events that should have had an effect on cost but are not revealed in the profiles. In addition, some profile changes may not be adequately explained by events in the time line. Such occurrences should be noted and combined with the events that were tagged during time line creation as being lacking an explanation. This composite list is a record of activities that the analyst needs to investigate further to be certain that all probable causes of cost growth are indeed the root causes. The analyst will then have the basic information needed to complete a narrative explaining how program activities led to the Nunn-McCurdy breach. Most core elements of the narrative should be evident after the matching exercise.

Reconcile Remaining Issues

At the end of the matching exercise, the analyst may have a list of activities that are still not fully explained. At this point, all available official sources will have been consulted. Secondary sources such as the trade literature can also be consulted. The analyst's list of unresolved issues can be used to initiate a round of interviews with program officials and with personnel in the PARCA office to determine if additional data sources can be consulted to gain insights into the issues that require further explanations.

Attribute Unit Cost Growth to Root Causes

During the construction of the unit cost profiles, the RCA team calculated the amount of cost growth attributable to each explanation provided by the SARs. The team may have found that the unit cost growth was caused by more than one event, that a single event might dominate, or that a combination of events may have together led to the Nunn-McCurdy breach.

In this activity, the RCA team should try to determine how much of the unit cost growth was due to each root cause. Since the root causes and chain of events that led to the Nunn-McCurdy breach can vary widely, there is no standard method for making such determinations. A generic approach is presented below to initiate the attribution analysis.

The RCA team can use the unit cost profiles to determine if there were any substantial changes to the unit cost before the Nunn-McCurdy breach. If so, then the explanations and decomposition of these substantial changes indicate that the program was experiencing significant problems before the breach, so the decomposition offers an indication of the nature of the root cause. For example, if the unit cost profile exercise shows that substantial unit cost changes just before the Nunn-McCurdy breach are attributable to the engineering and estimating categories, then the RCA team can focus on engineering issues in the program such as technical problems and on estimating issues such as a revised cost estimate. In this illustration, technical problems can be a root cause, but a revised estimate is not a root cause because cost estimates are revised for a reason. The RCA team needs to investigate the reason for the revised estimate. The team may discover that the cost estimate was revised because another test was added in response to technical problems. If so, then the root cause of the unit cost growth is technical problems, which the RCA team can describe in more detail. The result of this exercise should be a percentage partition (a ballpark estimate) of the unit cost growth with each part attributable to a root cause.

Create the Postulates and Deliverables and Reform the Original Hypothesis

All postulates and deliverables described below are submitted to the senior advisor for root cause analysis within the PARCA office. Because of the limited time available to perform the analysis, the RCA process to produce the postulates and deliverables is also used to progressively hone the original hypothesis used to guide the research. During production of the deliverables and postulates, the RCA team will gain greater insight into the nature of the program as new material is uncovered and then synthesized with existing matter from the program archives. In light of new findings, the original hypothesis used as the basis for the research effort is reevaluated and modified as appropriate. In this sense, an RCA process is iterative—new findings contribute directly to the various deliverables and postulates, as the research process advances the primary hypothesis simultaneously.

The typical set of deliverables and postulates includes a summary narrative that includes the completed PARCA office–generated root cause matrix, a root cause briefing that corresponds to the narrative, and a full root cause analysis report. The narrative and briefing are delivered well within the 60-day time frame to allow sufficient time for PARCA office personnel to use the RCA results in their submission to the Office of the Under Secretary of Defense (Acquisition, Technology, and Logistics) (OUSD(AT&L)). The full report can usually be submitted after the 60-day time frame. In addition, PARCA officials are given a copy of the data archive, and the data sources record is also available to them.

The Narrative

Deliverables will typically include a short narrative explaining the chain of events that led to the Nunn-McCurdy breach. This narrative should include formal documentation of the cost threshold breach. The narrative should also explicitly state the root causes of the unit cost growth and indicate how much of the unit cost growth is attributable to each root cause. A summary of the time line that includes the chain of events that led to the Nunn-McCurdy breach should be included in the narrative. In addition, the RDT&E, procurement, and unit cost profiles should accompany a summary of the results of the matching exercise among these profiles and the historical time line. Finally, a completed PARCA office–generated root cause matrix should be included with summary explanations for each cell that is marked as relevant. A description of the root cause matrix is in the following section.

The PARCA Office–Generated Root Cause Matrix

The PARCA office derived a root cause matrix from Title I § 103(d) of the WSARA. This matrix is shown in Table 2.2. The first column of the matrix lists eight issues that could lead to unit cost growth. Three of these are baseline issues:

- unrealistic estimates for cost or schedule
- immature technology, excessive manufacturing, integration risk
- unrealistic performance expectations.

The five listed below are execution issues:

- changes in procurement quantity
- inadequate funding/funding instability
- unanticipated design, engineering, manufacturing, or technology issues
- poor performance of government or contract personnel
- other.

The first row of the matrix should list the key events that led up to the Nunn-McCurdy breach along with the fiscal year of each key event.

The RCA activities conducted in the course of identifying the chain of events leading to the Nunn-McCurdy breach should allow the RCA team to complete the matrix. For each key event, the RCA team should identify the issue that arose during the key event that led to the breach. For example, if a quantity change occurred at Milestone (MS) B, then the cell corresponding to "Changes in procurement quantity" and "MS B" should be completed with a few words indicating the magnitude of the change, such as "quantity reduced 50 percent to 100 units."

The Root Cause Briefing

A briefing that corresponds to the narrative should be created for the OUSD(AT&L). This briefing should include the completed PARCA office–generated root cause matrix as well as a conclusions chart that attributes the unit cost growth to root causes.

The Full RCA Report

Although the narrative summarizes the RCA and conclusions, the full RCA report, which is a research report, will provide all of the details.[8] This report should include a description of the program, background information, and a description of the approach used to conduct the RCA. Data used in the RCA and data sources including any interviews should be identified and described. The historical time line should be presented. Each key event in the time line should be explained in full detail including its link to the chain of events that led to the Nunn-McCurdy breach. The RDT&E, procurement, unit cost, and program cost profiles should be shown with explanations for notable changes. Another set of graphics that connect the historical time line with the cost analysis should follow with detailed explanations of how the matches between key events and cost changes were made. A description of how the RCA team attributed

---

[8]   See, for example, Blickstein et al., 2011, and Blickstein et al., 2012.

Table 2.2
Root Cause Matrix Generated by the PARCA Office

	Year from MS B and Fiscal Year					
	B 2001	+1 2002	+2 2003	+3 2004	+4 2005	+5 2006
Baseline issues						
Unrealistic estimates for cost or schedule		X	X	X	X	X
Immature technology, excessive manufacturing, integration risk		X	X	X	X	X
Unrealistic performance expectations		X	X	X	X	X
Execution issues						
Changes in procurement quantity	X	Change from 150 to 55				
Inadequate funding/ funding instability	X					
Unanticipated design, engineering, manufacturing, or technical issues	X					
Poor performance of government or contract personnel	X					

unit cost growth to the root causes should follow. The PARCA office–generated root cause matrix should be presented along with descriptions for each cell identified as relevant. The report should end with a conclusions chapter that summarizes findings, notes limitations, and itemizes areas of continuing or future risk. All source material should be itemized in a reference list or bibliography.

CHAPTER THREE

Data Sources for Root Cause Analysis

To conduct a successful RCA, analysts require reliable and complete data derived from a large compilation of documentation on programs that are in Nunn-McCurdy breach. The sources of these data range from "official" program documentation to trade literature on various RCA programs. Analysts must understand a program's detailed history from a variety of viewpoints. These include the following perspectives:

- military requirement
- financial (cost and funding)
- technical
- contractual
- schedule
- acquisition environment.[1]

Different information and data are required to create the program history from these different viewpoints. The methodology section discussed how the analyst combines the compiled views to isolate the root cause(s) of a Nunn-McCurdy breach. Information and data for the RCAs conducted by RAND have been obtained from a variety of sources, but RAND's preference is to use primary sources and official documents to ensure the validity of the data.

A broad comprehension of the need that the program is intended to fulfill enables the analyst to interpret program decisions and events from a mission prospective. Hence, a program's history from initial concept and program inception to events leading to the Nunn-McCurdy breach is required background knowledge. Because the focus of the breach (or other large cost growth that invites high-level interest) is excessive cost growth, the analyst must be able to construct a financial view of the program in terms of how the program evolution was envisioned, the funding stream, and a time line of program cost estimates.

---

[1] While performing the first six RCAs, in addition to internal decisionmaking, RAND found that events/decisions external to the program may have major influences on individual programs. This was the case for the DDG-1000, WGS, and Excalibur programs.

In this chapter, we present and describe data documents typically used and consulted by the RAND RCA teams during the six root cause analyses completed to date. The chapter closes with a description of the archiving process RAND has created to preserve the data used and make it accessible for future PARCA office efforts. In Appendix A, we present these data in the aggregate, by source, level of restriction, and functional area within the RCA methodology. This allows us to display the relative level of access to some documents. It also gives us the opportunity to plan for long-lead-time activities by way of other sources.

Data Document Descriptions

Table 3.1 lists documents useful for conducting an RCA. Each RCA is unique, so a particular case may involve only some of the documents whereas another might involve a document that was not necessary for the first six RCAs. One key to a successful root cause analysis is not only researching what causes cost growth but also eliminating what did not cause cost growth. Generally, a scan through these sources will allow an RCA team to make eliminations and move on to what is at the core of the cost growth. The list in the table is meant to be representative and informative. Short descriptions of each type of document are listed in Table 3.1 along with typical document locations and a primary use of the data source based on RAND's RCA experience.[2] A more detailed description of each item follows the table.

Acquisition Program Baseline
The APB satisfies a statutory requirement for every program manager to document program goals before program initiation. The intent of the APB is to provide a set of programmatic, schedule, and financial constraints under which the program will be managed. Program goals consist of an objective value and a threshold value for each parameter. Objective values represent what the user desires and expects. Thresholds represent the acceptable limits to the parameter values that, in the user's judgment, still provide the needed capability. Failure to attain program thresholds places into question the overall affordability of the program or the capability provided by the system. This baseline document is a summary and does not provide detailed program requirements or content. However, it does contain key performance, schedule, and cost parameters that are the basis for satisfying an identified mission need.

Some of the information in the APB is provided in the SARs. This duplication of information allows analysts to both confirm the validity of the data and use the SARs primarily for APB information because SARs have a wealth of other information that

---

2   The PARCA office can help the RCA team obtain documents that are difficult to access.

Table 3.1
Typical Data Sources Used for RCAs

Name	Description	Source	Primary Use
*Acquisition strategy	Provides the program's plan for each portion of its life cycle from requirements to logistics and sustainment	Program office or OUSD(AT&L) or service-level acquisition executives (actual contact varies by program)	Construct time line of key events; key source for past information on the program
*ADM	Provides key decisions by Milestone Decision Authorities on programs	Program office or OUSD(AT&L) or service-level acquisition executives (actual contact varies by program)	Understand/identify key decisions made by OSD and the services regarding a program's direction; also used to understand a program's history and to create a time line of key events
APB	Establishes a baseline of programmatic, schedule, and financial constraints against which a program is managed	Defense Acquisition Management Information Retrieval (DAMIR, May 14, 2012)	Construct time line of key events
Budget Estimate Submission (BES)	Provides a program's funding summary, annual funding by appropriation, funding projection, and unit cost report	DAMIR	Verify cost and quantity data
Congressional testimony	Tracks "official" program office, service-level acquisition executive, and OUSD (AT&L) verbal exchanges on congressional concerns regarding programs	Government Printing Office, "Congressional Hearings," undated	Use depends on content
*Cost Analysis Requirements Description (CARD)	Up to three cost estimates may be prepared for an Acquisition Category ID program approaching a major milestone: the program life-cycle cost estimate (LCCE), an optional component cost analysis (CCA), and an independent cost estimate (ICE); all offices preparing estimates use CARD to ensure that the estimates are comparable	CAPE office	Construct cost history and understand cost issues

Table 3.1—Continued

Name	Description	Source	Primary Use
Congressional Budget Office (CBO) cost estimates	Provide another set of cost estimates for programs that can be compared with CAPE, service-level, and program office estimates	CBO	Construct cost history and understand cost issues
CRS reports	Provide independent assessments of program problems and issues over time; can be used for secondary validation of RCAs	Variety of websites	Use depends on content; generally used as a secondary assessment of the program's difficulties
*DAES	Provides a large amount of information on MDAPs in quarterly reports; types of information include program information (contacts), mission and description, summary of program to date, threshold breaches, schedule, performance, track to budget, cost and funding, low-rate initial production (LRIP), nuclear cost, foreign military sales, unit cost, contracts, deliveries and expenditures, operations and support (O&S) cost, program manager and OSD assessments, and sustainment	DAMIR	Time line of key events and cost history
DAES monthly program status charts	Provide a monthly program assessment of cost, schedule, performance, funding, and life-cycle sustaininment issues along with summaries of pressing issues, risks, technological performance, and interdependencies with other programs	DAMIR	Time line of key events and cost history, and lays out risks, challenges, and performance month by month; monthly details provide better understanding of challenges and what program is doing to fix identified challenges
*DCMA monthly reports	Provide in-depth analysis by DCMA on the performance of major contractors/contracts for MDAPs	DCMA, prime contractors, program offices	Time line of key events, cost history, and contractor performance
Earned value management (EVM) system data	Provide access to top-level EVM data on major contracts for MDAPs	DAMIR	Identifies cost and schedule problems at the highest level on largest contracts; monthly data allow tracking during a contract's performance

Table 3.1—Continued

Name	Description	Source	Primary Use
External studies/policy decisions affecting RCA program outcomes	Provide decisions and rationale behind changes in mission focus, etc., that affect program outcomes beyond the program office at the service or OSD levels	Service-level acquisition executives, program offices, OUSD(AT&L) (actual contact varies by program)	Use depends on content
Government and industry-wide collaboration sources	Provide access to award fee, cost estimates, Contractor Performance Assessment Reporting System (CPARS), design reviews, prime contractor contracts, and other program documents	Program offices, prime contractors (actual contact varies by program)	Time line of key events and cost history and identification of key challenges
*Interviews with program office personnel, contractors, CAPE, or service-level cost agencies	Provide current information on the state of the program through first-hand experience	Program offices, prime contractors, service-level cost agencies (actual contact varies by program)	Use depends on content; generally helps to understand a program's challenges and breaches
*Letters of notification to Congress of Nunn-McCurdy breach	Provide "official" rational to Congress for Nunn-McCurdy breach by service-level acquisition executive	Service-level acquisition executives, program offices, OUSD (AT&L) (actual contact varies by program)	Document Nunn-McCurdy breach
*Nunn-McCurdy OIPT cost and management briefings	Provide most up-to-date information on program to OUSD (AT&L) officials along with a plan to analyze and remedy problems; output of these meetings is justification of all statutory mandates for recertification	Service-level acquisition executives, program offices, OUSD (AT&L) (actual contact varies by program)	Time line of key events, cost history, and up-to-date information on a program's challenges
*Official briefings	Provide program status reports by the program office to various acquisition officials; discuss current problematic issues	Service-level acquisition executives, program offices, OUSD (AT&L) (actual contact varies by program)	Use depends on content; source for key events, cost/schedule history, contractor performance, technological challenges, risks
President's budget	Contains the budget message of the President, information on the President's priorities, budget overviews organized by agency, and summary tables; it is typically issued in February of the year before the start of the fiscal year	The White House, "Budget of the United States Government, Fiscal Year 2013," Washington, D.C.: Office of Management and Budget	President's priorities, budget overviews by agency, and summary budget tables to document time line

Table 3.1—Continued

Name	Description	Source	Primary Use
POM	Provides total acquisition cost and quantity summary, funding summary of appropriation and quantity, annual funding by appropriation, and unit cost report	DAMIR	Document Nunn-McCurdy breach
*Program Deviation Report (PDR)	"Official" memorandum from program manager to service-level acquisition executive that provides notice that the program is now in Nunn-McCurdy breach and the reason for the breach	Service-level acquisition executive, program offices, OUSD(AT&L) (actual contact varies by program)	Document Nunn-McCurdy breach
RAND and other studies	Prior FFRDC studies can provide general insight into cost growth and issue for particular MDAPs	For examples, see RAND's external website or that of the Institute for Defense Analyses	Use depends on content; program history/challenges are followed in some of these documents
*SARs	Like the DAES, these annual reports provide a wealth of data on the following topics: program information (contacts), mission and description, summary of program to date, threshold breaches, schedule, performance, tract to budget, cost and funding, LRIP, foreign military sales, nuclear cost, unit cost, cost variance, contracts, deliveries and expenditures, and O&S cost	DAMIR	Time line of key events; consolidate source of cost, quantity, contract, and schedule information
Trade literature	Provides linkages to missing details about programs because MDAPs are generally followed closely throughout their life cycles	Vendor websites available by subscription	Use depends on content
*Undersecretary of Defense for Acquisition, Technology & Logistics (OUSD(AT&L)) or service-level acquisition executive non–ADMs on program	Key documents that track senior-level decisions made regarding the program	Service-level acquisition executive, program offices, OUSD(AT&L) (actual contact varies by program)	Time line of key events, decisions, and cost history
GAO reports	Provide a secondary source that can be used to verify findings on problematic issues associated with a weapon program	GAO	Use depends on content; typically, GAO covers program challenges

NOTE: An asterisk by a program's name indicates that this document has been a "key" document for conducting RCAs.

is critical for understanding program performance. A key use of APBs is in constructing a time line of key events.

Acquisition Decision Memorandum

ADMs are key documents needed to conduct RCAs. The two basic purposes of an ADM at a major milestone are to record the decision made by the Defense Acquisition Executive (DAE) and to provide direction to the component program manager or other relevant party. The ADM package also includes any other documents that require DAE signature or approval, such as the APB or acquisition strategy. These documents are critical for providing analysts with an understanding of important decisions made at the highest levels of management. They can be used to construct a time line of relevant events and also to reconcile and highlight issues that are problematic for the program. Most important, these documents tie issues to unit cost growth when cost growth has become a major issue for the program—an issue that requires acquisition executive decisionmaking.

There are two potential difficulties with using these documents. The first is access and the second is their ad hoc nature. These documents are not readily available to the public and are ad hoc or released as needed, so analysts must rely on program offices to supply the entire set. Therefore, they are sometimes difficult to retrieve, and verification that all ADMs for a program have been collected may be difficult if not impossible to prove. The material in ADMs is often useful in constructing a time line of key events.

Acquisition Strategy

The acquisition strategy contains valuable information for RCAs. The acquisition strategy is prepared before the program initiation decision and updated before all major program decision points. Acquisition strategies are a rich source for information on various aspects of a program's life cycle. Some of the topics covered in the acquisition strategy include consideration of requirements, program structure, acquisition and contracting approach, risk management, program management, support strategy, and business strategy. Acquisition strategies are useful in RCAs because they provide a good source of historical events, source documents, and decisions that were made throughout the program. This is useful in constructing a time line of relevant events.

In one respect, they are more useful than SARs in constructing a time line for a program because the information is typically in one document, whereas analysts need to go through multiple SARs for MDAPs to create a time line. However, a drawback to using these documents is that they need to be provided by the program offices or acquisition executives, so their accessibility can be more limited than that of SARs. Having these data in DAMIR simplifies access because it does not require program office participation. In addition, there are typically multiple acquisition strategies depending on how many major shifts in program focus have occurred. This also limits accessibility, since only a person with knowledge of a program's background would know when these documents were approved. The acquisition strategy is useful for constructing a time line of key program events.

Budget Estimate Submission

The Budget Estimate Submission document is developed with each budget cycle and provides a program's funding summary, annual funding by appropriation, funding projection, and unit cost report. It is a summary of top-level budget data.

The BES is available through DAMIR, so access is limited to whether an analyst has access to DAMIR. Having these data in DAMIR simplifies access for RAND because it does not require program office participation in retrieving the document. In addition, anything from DAMIR is retrieved immediately at the start of each RCA, so there is no need to wait for these data. This document is not considered to be a key document for an RCA, but the data contained in a BES are vital for an RCA. This information is also provided in the yearly SARs and quarterly DAES reports, which have been used more often for the RCAs because they contain additional information on a variety of program topics. BES data can be useful in verifying cost and quantity data.

As budgeting processes change from administration to administration, some of the specific steps described above may change. However, the fundamental uses of budgetary information will remain constant.

Congressional Testimony

A compilation of congressional testimony on a program or external issues related to a program is important for understanding both the congressional and the services' stance on program problems. Congressional testimony tracks "official" verbal or written testimony between the program office, service-level acquisition executive, and OUSD(AT&L) and the Congress. In addition to testimony by acquisition executives, there is often testimony by other senior executive branch individuals including those charged with requirement setting. Congressional testimony can be used to identify reasons for cost growth, and the source is easy to access because the information is available to the public. Congressional testimony often has information useful for understanding how officials view the program. Use of testimony in an RCA depends on its content.

CARD, OSD/CAPE, and Service Cost Estimates

For Acquisition Category (ACAT) I and IA programs, the program manager must provide a final CARD to the CAPE office 45 days before the Milestone B OIPT. The DoD component Program Executive Officer (PEO) must approve the CARD. This is in accordance with DoD 5000.4-M,[3] which specifies CARD content. Also according to the instruction, the program described in the final CARD(s) at Milestone B is sup-

---

[3]   DoD 5000.4-M, "Cost Analysis Guidance and Procedures," December 11, 1992.

posed to reflect the program definition achieved during the technology development phase.

Up to three cost estimates may be prepared for an ACAT ID program approaching a major milestone: the program LCCE, an optional component cost analysis, and an ICE. All of the offices preparing estimates use the CARD to ensure that the estimates are comparable.

The CARD describes the program in sufficient detail to enable cost analysts to develop cost estimates for the program. Therefore, the program office, the service, and the acquisition community define the program in a considerable amount of detail—detail that they would have not developed unless this was a requirement. The CARD is a key document for conducting a root cause analysis because it helps to create a scenario describing whether the program is affordable over time by using a lot of detail about the program that cannot be found elsewhere. The cost estimates can also be checked for consistency with the CARD. However, these documents are very difficult to access. They must be retrieved from the CAPE through the PARCA office, and they contain sensitive cost data. The CARD is useful for constructing the cost history of the program.

Congressional Budget Office Cost Estimates
The Congressional Budget Office provides cost estimates to Congress that can be compared to other cost estimates from CAPE and the services. These estimates help to verify cost data. They are particularly important in cases where the CAPE and service-level estimates are not in agreement. CAPE will often issue an explanatory memo documenting its view of the cost differences. Cost estimates from the CBO are useful in constructing the cost history of the program.

Congressional Research Service Reports
CRS reports provide independent assessments of program problems and issues over time. These assessments are specifically made for Congress. The reports offer Congress another perspective on program problems. These reports can be used by analysts to validate findings in RCAs or to reconcile any outstanding issues that exist with a program. The only drawback is that they are sometimes difficult to retrieve because they are not readily available to the public. Copies can be requested through a member of Congress. The content of CRS reports differs and their use in RCAs depends on the topics addressed in specific reports.

Defense Acquisition Executive Summary
The information presented in quarterly DAES reports is critical for doing an RCA. DAES reports provide the following categories of information on a program: program information (contacts), mission and description, summary of the program to date, threshold breaches, schedule, performance, track to budget, cost and funding, low

rate initial production, nuclear cost, foreign military sales, unit cost, contracts, deliveries and expenditures, operating and support cost, program manager and OSD assessments, and sustainment.

DAES reports have very similar data to SARs, which are explained later in this section. Analysts thus far have used DAES as a way to bridge the gap in data from the most recent SAR to the Nunn-McCurdy breach. The data in the DAES reports can be used for a variety of purposes in a root cause analysis: documenting the unit cost threshold breach, constructing a time line of the program's history and pulling out relevant events, creating program cost profiles, matching a time line of events with cost profiles, reconciling remaining issues, and attributing unit cost growth to root causes.

DAES, like SARs, are available in DAMIR, so access at the beginning of the project is possible. Data in the DAES are not as rigorously verified as those in the SAR, so it is better for analysts to use the SAR as the primary source and DAES as a secondary. The DAES can contain material useful for constructing a time line of key events and understanding the cost history of the program.

Defense Acquisition Executive Summary Monthly Program Status Charts

DAES monthly program status charts are summary Microsoft PowerPoint slides that provide a snapshot of the following program indicators: cost, schedule, performance, funding, and life-cycle sustainment issues along with summaries of pressing issues, risks, technological performance, and interdependencies with other programs. They are useful because they provide detail on issues that the program is trying to resolve from month to month. These data are unique and cannot be found elsewhere. This source provides nuggets of information that may signal cost or schedule problems.

The format of these charts is difficult to analyze because dozens must be analyzed carefully to understand a program. The DAES charts are available through DAMIR and so are easily accessible at the beginning of the project and throughout for updates. The DAES and monthly program status charts offer a way to quickly review key program events. The material contained in the charts can facilitate creating a time line of key events and understanding the program's cost history.

Defense Contract Management Agency Monthly Reports

As a way for DoD to monitor contractor performance, DCMA provides monthly reports that have in-depth analysis of the performance of major contractors/contracts for MDAPs. They focus not only on EVM performance data but also on technical risks. This type of information is useful because it helps to determine whether a program is experiencing contractor performance issues or major technical barriers or risks.[4] In the case of one of the first five RCAs, poor contractor performance did factor into the cost growth.

---

[4] See Chapter Four in Blickstein et al., 2012, where this issue is discussed in greater detail.

These reports are not readily available through DAMIR, so FFRDCs such as RAND have to obtain them through the PARCA office. This limits ease of accessibility. They are also monthly, which is positive in that the data are current, but it is hard to draw conclusions from dozens of these reports. The DCMA monthly reports can offer details that help analysts understand a program cost history and identify key events for constructing a time line.

Earned Value Management System Data

EVM data are present in DAMIR. The data are at the top level only for major cost-type development contracts. They indicate whether contractor performance is an issue for a program. They are difficult to download from DAMIR in a format that can be manipulated; however, the data are monthly and usually current, so analysts investigating RCAs might find it useful to consult this information to begin to understand contractor performance and any relationship it may have with root causes and unit cost growth. This knowledge can be useful in constructing a time line of key events as well as in understanding a program's cost history.

External Studies/Policy Decisions Affecting RCA Program Outcomes

Two of the RCAs already conducted—Excalibur and DDG-1000—found that decisions at the service level but outside the program office had major effects on program outcomes. In both cases, cost growth was directly attributed to decreases in quantity based on the Navy's change to the service-level mission focus and the Army's change to the overall munitions profile. Analysts found that it is important to look for external factors as root causes of the program's cost growth. These decisions are typically recorded in studies done at the service level. However, they are very difficult to obtain because of the sensitivity of the decisions being made, and it is not always possible to obtain them. In the cases mentioned, we found that other sources—such as PDRs or ADMs—may provide a general level of understanding regarding service-level decisions. The specific use of external studies depends, of course, on the topics addressed by the studies.

Government and Industry-Wide Collaboration Sources

Access to government and industry-wide collaboration sources such as award fees, cost estimates, CPARS, design reviews, prime contractor contracts, and program documents has aided in the RCAs. These types of documents provide the perspective of industry. Depending on a program's underlying problems, these documents may be useful. They can help the analyst understand the key events for the program, verify cost data, reconcile remaining issues, and attribute unit cost growth to root causes. In the first five RCAs, several RCA teams had good access to industry. One team did not have as much access because of time constraints. Access is a main issue in working with industry. Contractors may or may not be willing to share sensitive information. RCAs found that in these cases, it is better to retrieve these data with help from the PARCA office. Knowledge of a contractor's perspective can be invaluable to understanding the procurement environment. Key program events may be identified with this knowledge.

Interviews with Program Office Personnel, Contractors, CAPE, or Service-Level Cost Agencies

Interviews with program office personnel, contractors, CAPE, or service-level cost agencies are a direct way to gather information from program and contractor personnel. These interviews can be extremely useful because they typically yield important, current insight into issues affecting programs. However, it is possible to set up only a few of these meetings in the short time frame allowed by the Nunn-McCurdy process. Interviews also require cooperation from the interviewees; program officials, contractors, and other personnel must be willing to address questions raised by the RCA and give priority to talking to an RCA team so that interviews can be conducted in a timely fashion. In this case, the content of the interview determines how the data are used in an RCA.

Letters of Notification to Congress of Nunn-McCurdy Breach

When a program experiences a Nunn-McCurdy breach, the secretary of the department concerned must provide a letter to the following offices in Congress with "official" rationale for the breach: the President of the Senate, the Speaker of the House, and the Committees on Appropriations and Armed Services in both the House of Representatives and the Senate. These letters constitute one of the first steps in the Nunn-McCurdy process. They are typically available through the PARCA office and they provide a quick understanding of what has happened to trigger the breaches, making them very useful for documenting the breach.

Nunn-McCurdy OIPT Cost and Management Briefings

The Nunn-McCurdy process is completed by the OUSD(AT&L) with a document recertifying or not recertifying a program. The OUSD(AT&L) has a statutory requirement to provide justification of various categories of information to complete recertification of a program. The justification for recertification comes from a series of briefings at the OIPT/Integrated Product Team (IPT) levels during the 60-day Nunn-McCurdy process. The following sources of information are consulted by the IPT team: site visit(s) to the program office/contractor, SAR and DAES inputs and programmatic documentation, and follow-up questions/interviews. After examining all the details, each IPT working group comes up with actionable recommendations for each deficient area to present to IPT leadership. Findings and final recommendations will be the basis of certification input and resulting ADM to Congress. These briefings are particularly useful for an RCA because they help to identify key parties involved in the Nunn-

McCurdy process and also root causes; however, they can be accessed only through the PARCA office.

Official Briefings

Official briefings offer a unique perspective on program status because they are generally given by the program manager or those close in rank to the program manager. The audience tends to be various acquisition officials including service-level acquisition executives and officials in OUSD(AT&L). These briefings are important to RCAs for several reasons: They provide first-hand information from the program office, they discuss pressing issues for the program, they are generally more current than other documentation, they are ongoing, and they offer a look at the program over time. Unfortunately, they are very difficult to retrieve because they have to come directly from the program office. They are also ad hoc, so it is difficult to know if all the briefings were given to the RCA analysts. Finally, they typically contain sensitive information and program analysis. Because the briefings are often sensitive, program managers may be reluctant to provide them to an RCA team. The specific content of these briefings determines their contribution to an RCA.

President's Budget

The President's Budget contains the budget message of the President, information on the President's priorities, budget overviews organized by agency, and summary tables. It is typically issued in February of the year before the start of the fiscal year. (For example, President's Budget 13 will be issued in February 2012.) More detailed budget material is generated by the services and is available as supplementary volumes on the services' financial websites.

Program Objective Memorandum

The POM is accessible through DAMIR and is a "current" statement of program funding/quantity status from the military department or defense agency. It provides the following information: total acquisition cost and quantity summary, funding summary of appropriation and quantity, annual funding by appropriation, and unit cost report. The information is important because it provides information on whether the program is having funding issues; however, most of these data are also in the SARs and DAES. The unit cost report helps the analyst derive the speeding ticket that is needed for the final analysis of a program.

Program Deviation Report

The PDR is an official memorandum from the program manager to the service-level acquisition executive that provides notice that the program is in Nunn-McCurdy breach and the reasons for the breach. This is accessible early in the analysis through the PARCA office or the program office and provides the clues to an initial understanding of the major causes of the breach that need to be verified through the RCA. In addition, this report explains why the program is important and also lays out the steps that the program office will take to continue the program while the Nunn-McCurdy process is taking place. PDRs are also useful for documenting the Nunn-McCurdy breach.

RAND and Other Studies

RAND and similar FFRDCs have a long history of tracking acquisition topics and major weapon system programs. All studies commissioned on the program under study, including but not limited to prior FFRDC research, can be used as a potential source to verify and identify programmatic issues. For example, RAND's history in tracking the DDG-1000 program was useful for conducting an RCA under a very tight deadline.

Selected Acquisition Reports

The information presented in annual SARs to Congress is critical for doing an RCA. SARs are congressionally mandated documents with information that has been rigorously verified. SARs summarize the following information on a program: program contact information, mission and description, summary of the program to date, threshold breaches, schedule, performance, track to budget, cost and funding, LRIP, foreign military sales, nuclear cost, unit cost, cost variance, contracts, deliveries and expenditures, and O&S cost.

SAR data can be used for a variety of purposes in a root cause analysis: documenting the unit cost threshold breach, constructing a time line of the program's history and pulling out relevant events, creating program cost profiles, matching a time line of events with cost profiles, reconciling remaining issues, and attributing unit cost growth to root causes.

SARs were one of the preferred sources used in all of the initial RCAs for a variety of reasons. The first is that they are in DAMIR, so access at the beginning of the project was immediate. Also, they provide a comprehensive program history from year to year along with a discussion of critical issues. In most cases, it was possible to start to understand root causes based on the cost variance explanations provided in each annual SAR. Data in SARs have also been rigorously verified because these documents are given to Congress. The validity of the data makes SARs a preferred source.

Although there are several minor drawbacks to using SARs, they do not outweigh the major benefits. The first is that SARs are released only annually unless a major problem in the program requires a quarterly update. To get the most up-to-date information on a program, analysts must consult other sources such as program briefings and DAES reports. Also, SARs must be read chronologically to piece together a time line because the SAR for each year is a separate document. This format slows down the ability to create a time line and track changes from year to year in cost, schedule, and threshold breaches. Finally, SARs can have classified sections, but in such a case, an unclassified version is usually available.

Trade Literature

A scan through the trade literature, such as *Inside Defense.com* or *Jane's*, is done after other sources have been exhausted.[5] This type of search can sometimes provide context for why certain decisions were made in the program. It may identify other events going on that are related or it might yield some understanding about a troublesome area. It can also reinforce some of the conclusions about the programs and reconcile remaining issues because MDAPs are covered heavily in the trade literature over the course of program life cycles. In the absence of some official sources, trade literature can be used with some caution. Specific use depends on the content of the literature.

OUSD(AT&L) or Service-Level Acquisition Executive Non-ADMs on a Program

Non-ADMs by OUSD(AT&L) or other service-level acquisition executives are highly useful documents for an RCA. They help to identify a critical set of decisions that have been made regarding the program over time. Like ADMs, they provide valuable insight into major problems that have justified an acquisition executive's input regarding the future of the program. These documents are not readily available and are ad hoc or released as needed, so analysts must rely on program offices to supply the entire set. Obtaining these documents can therefore be difficult. The information contained in these memos is useful for creating a time line of key program events and understanding the program's cost history.

Government Accountability Office Reports

The GAO examines weapon systems if requested by Congress. GAO reports can be generic or can examine a specific dimension of a weapon system's acquisition. Congress has generally requested GAO examinations when a weapon system exhibits such problems as noticeable cost growth, schedule slippage, or major technical barriers. GAO reports are used to inform Congress about the nature and details of problematic areas in weapon system acquisitions and hence are very useful for gaining insights into whether problematic areas can be linked to a Nunn-McCurdy breach. In particular, GAO reports can help construct a time line of relevant events, reconcile outstanding issues, and attribute unit cost growth to root causes. GAO reports are a secondary source that should be consulted after official program documentation has been reviewed. How useful GAO reports are in a particular RCA depends on the topics addressed in the reports.

---

[5] See also Chapter Four of Blickstein et al., 2012.

RCA Document Archive Process

Root cause analyses are ongoing. Consequently, an established methodology is needed that can be followed in future studies along with an accurate archive of sources used in past RCAs. A permanent archive has three purposes: preserving sources used in conducting previous RCAs, using prior research and sources to assist with future RCAs, and maintaining sources for future PARCA office efforts including analysis of the programs after recertification.[6]

RAND has established an archive that contains documentation from the first six RCAs. It is archived on a RAND RCA team SharePoint site where team members and PARCA office personnel scan access files as needed. The full list of sources used to inform the first six RCAs is presented in Appendix B. The archiving process is outlined below and illustrated in Figure 3.1.

The steps for archiving RCA documents are:

1. Archivist collects all useful documents from RCA team members and identifies documents that are considered to be "key" to the final analysis.
2. Archivist uploads electronic versions of collected documents to the current SharePoint site under folder names "ARCHIVE OF PRIOR RCAs (For RAND and PARCA Office use)" with subfolders relating to each program. At this point, anyone from RAND working on PARCA office–directed work would be able to access these files as long as they are given permission to join the SharePoint site.
3. RAND creates an external user account for PARCA office personnel if requested. PARCA office personnel desiring access will need to sign an agreement with RAND regarding external access to the RAND network. Access can also be provided via portable media such as compact disks.

Figure 3.1
RCA Document Archiving Process

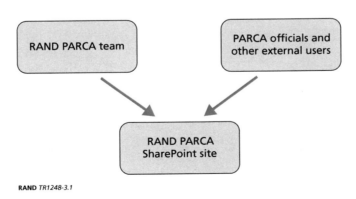

RAND *TR1248-3.1*

---

6   The WSARA charges PARCA with performing periodic performance assessments on MDAPs including programs that have received recertification after a Nunn-McCurdy breach.

4. Archivist creates and updates bibliographic entries of all collected documents.
5. Each bibliographic entry has a hyperlink attached to it. Clicking on the link takes the user to the SharePoint where he or she logs in. After logging in, the document is displayed. PARCA office personnel are given access to SharePoint directly to facilitate downloading of multiple documents.

Framework for Data Sources Database

In addition to creating an archive of the data sources used in RCAs, future researchers may need ways to search those data sources. For this reason, we have also created a searchable database of RCA sources (see Table 3.2). The database currently exists as a Microsoft Excel spreadsheet. In this form, it is intended as a prototype.[7] As we perform more RCAs, other formats for storing a searchable data sources database may prove more useful. Hence, in this report, we focus on describing a framework for a searchable RCA data sources database.

Our experience to date indicates that there are a number of useful search dimensions. The dimensions in our current framework are listed below.

- **Program name**

  The program name is the program RCA where the document was used.

- **Releasability constraints**

  Releasability constraints are the distribution constraints on the document at the time the document was used on an RCA. Security classifications such as "For Official Use Only" are examples of releasability constraints.

- **Author**

  The author is the name of the author of the document.

- **Title**

  The title is the official title of the document.

- **Organization**

  Organization is the name of the entity affiliated with the document. Examples include OUSD(AT&L) and the U.S. GAO.

---

[7] Microsoft Excel is often used as a platform for prototype exploration and proof of concept. Examples include the Electronic Decision Enhancement Leverager Plus Integrator (E-DEL+I, ©, ™), the Electronic Policy Improvement Capability (EPIC, ©, ™), and the government's JIC. After proof of concept, some developments migrate to more fitting platforms and others remain as Microsoft Excel–based capabilities. Endnote and Access are candidate migration platforms for the Root Cause Analysis Source Data Records Database.

Table 3.2
Searchable Data Sources Database Framework

Program	Author	Title	Organization	Repository	Source Identification Number
Apache Longbow (AB3)	Ahern, David G.	Memorandum for OUSD(AT&L): Apache Block III (AB3) OIPT Report	OUSD(AT&L) Portfolio Systems Acquisition		
Apache Longbow (AB3)	Ahern, Crosby, and Openshaw	Defense Acquisition Board: AB3A Remanufacture and AB3B New Build	OUSD(AT&L)		
DDG-1000		OSD CAPE Visit DDG-1000 Program Update	General Dynamics Bath Iron Works		
Excalibur	Bertuca, Tony	Army Using HTS Precision Munitions Biometrics to "Win Hearts and Minds"	Inside the Army		
Excalibur	Bolton Jr., Claude M.	Memorandum for Program Executive Officer Ammunition: Acquisition Decision Memorandum— Milestone (MS) C Decision for Excalibur XM982 Block Ia-1	Department of the Army Office of the Assistant Secretary of the Army (Acquisition Logistics and Technology)		
JSF		Joint Strike Fighter: Significant Challenges and Decisions Ahead	GAO		GAO-10-478T
JSF		Appendix H Risk	Joint Strike Fighter Program Office and Lockheed Martin		TR 7101-026A/09
Navy ERP		Navy Enterprise Resource Planning (ERP)	OUSD(AT&L)	DAMIR MAIS	

- **Repository**

  The repository is the name of the document repository where the document was found during an RCA.

- **Source identification number**

  The source identification number is the document identification number.

Category	Date	Keywords	Internet Link	Hyperlink	Hyperlink	Hyperlink	Hyperlink	Hyperlink	Notes
Report	8/30/10			AB3 OIPT Report					
DAB	9/27/10			AB3 OSD DAB Slides					
Visit	3/22/10			OSD Cape Visit 03222010.pdf					
Article	4/05/10			Excalibur trade literature from InsideDefense.doc					
ADM	5/23/05			Excalibur ADM 05232005.pdf					
Report	March 2010			d10478thigh.pdf					
Report	7/31/01			GE Govt Assess Risks applying to LM.doc	LM Govt Assess Risk (MCR).doc	LM Govt Risk Reports.doc	LM Risk assessment table.xls	PW Govt Assess Risks applying to LM.doc	
Report	Approval date: 6/30/10			NAVY ERP MAIS Quarterly Report (MQR) 06302010.pdf					

- **Category**

  The category is the class of documents to which the data source belongs. For example, all SARs are in the SARs category whereas all studies are in the Reports category.

- **Date**

  The date is the date that appears on the document.

- **Keywords**

  The keywords are words that describe the contents of the document. Examples of keywords include "cost analysis" and "ship technologies."

- **Internet link**

  The Internet link is an active Internet address for the document.

- **Hyperlink**

  The hyperlink is an active link to the data sources archive described in the preceding section.

- **Notes**

  Notes are any comments that researchers wish to add to further illuminate the contents of use of the document.

CHAPTER FOUR
Conclusions

RAND's experience in conducting six RCAs indicates that each acquisition program is unique, and each RCA is unique. That said, RAND's experience also indicates that a set of core activities is instrumental to a successful root cause analysis. These activities constitute a generic root cause methodology whose key components include the following:

- Gather and review readily available data.
- Develop a hypothesis.
- Set up long-lead-time activities.
- Document the unit cost threshold breach.
- Construct a time line of cost growth relevant events in the program history.
- Verify the cost data and quantify cost growth.
- Create and analyze the program cost profiles pinpointing occurrences cost growth.
- Match the time line events with the changes in the cost profiles and derive root causes of cost growth.
- Reconcile any remaining issues.
- Attribute unit cost growth to root causes.

Carrying out this set of activities should enable the research team to create the primary deliverables and postulates for a root cause analysis: a summary narrative that includes clearly stated root causes of cost growth supported by a formal documentation of the cost threshold breach, a summary time line of program events leading to the Nunn-McCurdy breach, funding profiles, a completed PARCA office–generated root cause matrix, and a breakdown of the amount of cost growth attributable to each root cause; a briefing that corresponds to the narrative; and a full root cause report.

In addition to developing deliverables and postulates, the RCA process is designed to improve the research focus iteratively. At each stage of the RCA, information is both drawn from and contributed to the program archive. The RCA analytic team uses insights from this information not only to improve the interim products that result from successive stages of the RCA but also to advance the original hypothesis that guides the research. This process of regularly refining the guiding hypothesis with the insights gained during the production of key deliverables and postulates enables the research teams to identify quickly the root causes of a program's failure.

Assessment of Data, by Provider and Functional Area

The Value of Data Early in the Analysis

The abbreviated amount of time available to perform root cause analysis places a high value on having data available both early and throughout the analysis period. Early availability of data is important, as it becomes the basis for establishment of the initial hypothesis indicated in the process for conducting root cause analysis described in Chapter Two. Figure 2.1 reflects the central importance of the hypothesis both at the beginning and at the end of the effort. Without early access to data, it becomes most difficult to arrive at that initiating hypothesis.

Just as having data early is important, having access to data from a variety of sources is also important so that the initiating hypothesis is not biased by limited data source perspectives. Appendixes A–F provide a rich depiction of the data used and the approaches to their collection. It is important that there be a clear understanding of the nature of the bibliographical collection in those appendixes so that the planning of early activity addressed in Chapter Two can be more systematically conducted.

This appendix arrays data and data sources in a way that enables the systematic planning activity necessary. To simplify the portrayal of the data, the appendix uses the very rich bibliography attendant on the *Zumwalt*–class Destroyer DDG-1000 root cause analysis to establish the parameters of the nature of the bibliography.[1]

DDG-1000 as an Example

In evaluating the six defense programs assigned to date, RAND reviewed a total of 845 source documents. In aggregation, these data point to a few patterns related to the source and functional area associated with the documents and programs.

Analysis of the DDG-1000 was based on 135 source documents, 82 (61 percent) of which were marked as FOUO, two (1 percent) marked as company proprietary, two (2 percent) marked for government use only, and the remainder marked unclassified or

---

[1] The DDG-1000 root cause analysis begins on p. 44 of Blickstein et al., 2011.

without other designation. The distribution of sources by classification for the DDG-1000 is not much different from that of the other programs reviewed, with the exception of the Navy ERP (13 percent of the documents were marked official unpublished government documents or company proprietary). The distribution of documents by classification is important in that access to restricted data would be difficult to attain before the breach and the formal initiation of a Nunn-McCurdy root cause analysis. In our example, although some data may have been available to RAND before the breach, the majority of the material would not have been available until after the investigation began. Figure A.1 shows the distribution of data by source restriction level for all programs, which is generally appropriate for the six analyses assigned and as well as for our specific example: the DDG-1000.

The source of the information also plays an important role in determining whether there will be ready access to data before or very early at the start of the analytic effort. Figure A.2 depicts the sources of data used during the DDG-1000 root cause analysis. Data sources such as GAO, CRS, CBO, and RAND are openly accessible and provide material early in the process. For the remainder of data sources, the availability of the data before the formal initiation of a Nunn-McCurdy root cause analysis is governed by the nature of the data release by the respective organization.

By organization, the majority of the data was derived from OUSD(AT&L), mostly in the form of SARs. Many of these data are available through access to the DAMIR online portal, even though restricted to DoD Common Access Card holders with at least DAES- and SAR-level access.

Figure A.1
All Program Data Sources, by Restriction Level

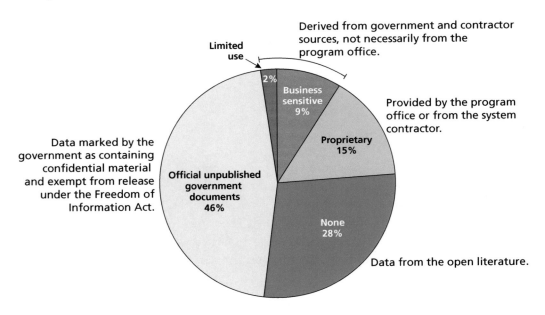

Figure A.2
DDG-1000 Sources, by Organization

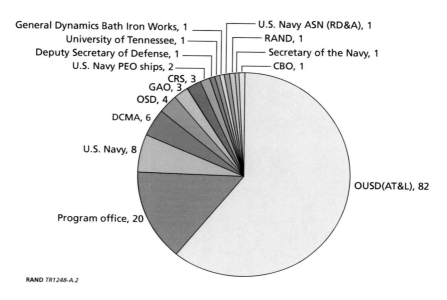

RAND *TR1248-A.2*

Because OUSD(AT&L) in the person of PARCA is the tasking agent for the analysis, it can facilitate making data available by providing access to OSD-controlled data systems.

The next largest source organization was the program office, followed by the U.S. Navy, DCMA, and OSD, which were sources for a few documents. Access to data from these sources is difficult to obtain before formal initiation of the Nunn-McCurdy root cause analysis.

Another depiction that characterizes the source of information and therefore its availability is found in Figure A.3. Here, the source categories used were OSD, program office, Congress, and external. As can be seen, the predominant sources of information were OSD followed by the program office. This reinforces the need to have clear DoD protocols governing the availability of information.

To gain further perspective, the documents were also categorized by their functional contribution to the analysis discussed in the methods description outlined in Chapter Two. These categories include acquisition, communications, contractual, and finance material. The majority of the documents fall into the acquisition document category (see Figure A.4). The next largest category, communications, included letters, statements, memos, and non-required Powerpoint presentations and briefings. Contractual documents included the APBs as well as required testimony or reports by the program office. The few finance documents were derived from Navy PEO ships and OUSD(AT&L).

Figure A.3
DDG-1000 Documents, by Source Categories

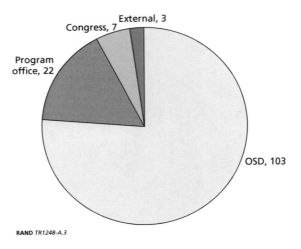

RAND *TR1248-A.3*

Figure A.4
DDG-1000 Documents, by Functional Area

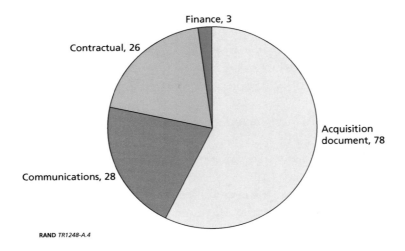

RAND *TR1248-A.4*

This depiction reinforces the need for well-established protocols for data access. Although the communications category is substantial and can provide some initial information for developing a hypothesis, the acquisition, contractual, and finance functional areas, which constitute three-quarters of the data sources used, represent problematic access controls.

To support planning efforts, analysts must understand the nature and source of the data and other documented information and also the production time line of material used in the analysis. As can be seen in Figure A.5, most documents reviewed were produced in the three-year period before the occurrence of the breach.

It is important to note that some of the OSD documents were available as early as 1998, whereas the other document types (external, congressional, and program office) were published between 2004 and 2010, the majority between 2007 and 2008. However, because of the classification and proprietary nature of much of the material, the complete data collection and analysis would not have been possible before a Nunn-McCurdy breach investigation. This suggests that once the appropriate material is made available, the research focus can reach back for a substantial period of time. The availability of historical data would contribute to a fuller expansion of the initial hypothesis addressed earlier.

This review of the sources by time, functional area, and organization suggests that access to OSD data sources is the most productive path—one that would allow the analyst to accumulate the majority of the data early on. Given the classification, proprietary, and business sensitive nature of some of this material, protocols are needed to govern the review of the data before as well as after a formal investigation of the program is under way. Obviously, public literature will be available before a breach, notably from providers such as the GAO, CRS, and CBO. Review of these public documents will be a component of any long-lead-time activities before a breach.

Figure A.5
DDG-1000 Documents, by Year Published

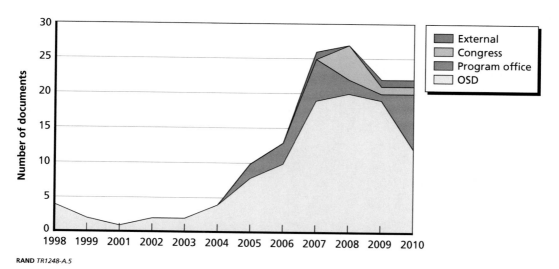

Understanding the Bibliographies

Development of an initial hypothesis is necessary for the root cause analysis of any platform. Without early access to a program's functional data elements—finance, acquisition, contractual, and communications—crafting the hypothesis can be difficult. As depicted by the DDG-1000 example, the majority of functional information was in the form of acquisition documents, many of which were available to analysts with the appropriate access level through DoD acquisitions resource portals, such as DAMIR, well before the breach. Therefore, early communication with the PARCA office is important to identify programs of interest and to start building an expertise even before the Nunn-McCurdy breach occurs. This early thinking could also help structure post-breach requests for data from the program office, OSD, and other major sources. These early-lead-time activities will reduce the strain to develop an initiating hypothesis and to perform root cause analyses in the future.

After a breach investigation has been initiated, analysts will have access to company proprietary, business sensitive, government classified, and FOUO data that otherwise are unavailable to researchers without an immediate need-to-know. Once granted this higher-level access to the program data, it is important that the analyst capture information from a wide variety of sources so as not to bias the hypothesis toward any one perspective. The DDG-1000 data from the program office were balanced by data from congressional, external, and OSD sources. These sources include data from several research organizations such as CBO, CRS, GAO, and RAND as well as DCMA and OUSD(AT&L) that provide valuable information about program risk to meeting the requirements. To build a strong initiating hypothesis a variety of sources are needed to balance the various perspectives underlying each piece of data.

Understanding the root cause of a Nunn-McCurdy breach requires a multifaceted approach to culling the data that should begin as early as possible. Early conversations with the PARCA office about programs that may be of interest would give an analyst the opportunity to begin crafting an initiating hypothesis from available sources before a breach. Once access is granted to any remaining restricted data, care should be taken to balance the information by surveying a variety of sources so as not to introduce any bias. The root cause analysis of a Nunn-McCurdy breach is a challenging undertaking that needs to begin early with care taken to the nature of the information being gathered.

Data Sources for the Longbow Apache Block III (AB3)

Note: an asterisk indicates a key source used in the Root Cause Analysis.

*"AB3," *Office of the Under Secretary of Defense (Acquisition, Technology & Logistics),* Document Acquisition Management Information Retrieval (DAMIR), Selected Acquisition Report (SAR), December 31, 2006.

*"AB3," *Office of the Under Secretary of Defense (Acquisition, Technology & Logistics),* Document Acquisition Management Information Retrieval (DAMIR), Selected Acquisition Report (SAR), December 31, 2007.

*"AB3," *Office of the Under Secretary of Defense (Acquisition, Technology & Logistics),* Document Acquisition Management Information Retrieval (DAMIR), PB10 Limited Selected Acquisition Report (SAR), December 31, 2008.

*"AB3," *Office of the Under Secretary of Defense (Acquisition, Technology & Logistics),* Document Acquisition Management Information Retrieval (DAMIR), Selected Acquisition Report (SAR), December 31, 2009.

*"AH-64D Longbow Apache Block III, DRAFT Procurement Objective Cost Analysis Requirements Description (CARD)," *US Army, Project Manager Apache,* February 2010.

*Bailey, LTC George D., "Probability of Program Success Summary," *US Army, Product Manager AB3,* June 8, 2010.

*"Longbow Apache," *Office of the Under Secretary of Defense (Acquisition, Technology & Logistics),* Document Acquisition Management Information Retrieval (DAMIR), Selected Acquisition Report (SAR), December 31, 2009.

*"Modernizing the Army's Rotary-Wing Aviation Fleet," *Congressional Budget Office (CBO),* November 2007.

*Openshaw, COL Shane T., "Memorandum Thru: Program Executive Officer Aviation; For Under Secretary of Defense (Acquisition, Technology, and Logistics): Longbow Apache Block III Program Deviation Report," *US Army, Project Manager, Aviation,* March 29, 2010.

*"PARCA: AB3 Questions for the Army," *RAND Corporation,* April 21, 2010.

*"PARCA: AB3 Questions for the Army: Block II," *RAND Corporation,* April 27, 2010.

*"PARCA: AB3 Questions for the Army: Block II: Revised," *RAND Corporation,* May 6, 2010.

*Rodrigues, Louis J., "Letter to: The Honorable William S. Cohen, The Secretary of Defense," *United States Government Accountability Office (GAO),* B-275846, January 27, 1997.

Data Sources for the DDG-1000

Note: An asterisk indicates a key source used in the Root Cause Analysis.

*Christian, John, "DRAFT Shipbuilding Programs Acquisition Framework," *OUSD(AT&L)/A&T/PSA/NW*, October 10, 2007.

"DDG 1000," *Office of the Under Secretary of Defense (Acquisition, Technology & Logistics),* Document Acquisition Management Information Retrieval (DAMIR), Acquisition Program Baseline (APB), Approval Date: January 12, 1998.

"DDG 1000," *Office of the Under Secretary of Defense (Acquisition, Technology & Logistics),* Document Acquisition Management Information Retrieval (DAMIR), Acquisition Program Baseline (APB), Approval Date: March 11, 1999.

"DDG 1000," *Office of the Under Secretary of Defense (Acquisition, Technology & Logistics),* Document Acquisition Management Information Retrieval (DAMIR), Acquisition Program Baseline (APB), Approval Date: April 23, 2002.

"DDG 1000," *Office of the Under Secretary of Defense (Acquisition, Technology & Logistics),* Document Acquisition Management Information Retrieval (DAMIR), Acquisition Program Baseline (APB), Approval Date: April 14, 2004.

"DDG 1000," *Office of the Under Secretary of Defense (Acquisition, Technology & Logistics),* Document Acquisition Management Information Retrieval (DAMIR), Acquisition Program Baseline (APB), Approval Date: November 23, 2005.

"DDG 1000," *Office of the Under Secretary of Defense (Acquisition, Technology & Logistics),* Document Acquisition Management Information Retrieval (DAMIR), SAR Baseline, January 13, 1998.

"DDG 1000," *Office of the Under Secretary of Defense (Acquisition, Technology & Logistics),* Document Acquisition Management Information Retrieval (DAMIR), SAR Baseline, January 25, 2006.

*"DDG 1000," *Office of the Under Secretary of Defense (Acquisition, Technology & Logistics),* Document Acquisition Management Information Retrieval (DAMIR),

Selected Acquisition Report (SAR) RCS: DD-A&T(Q&A)823-197, December 31, 1998.

*"DDG 1000," *Office of the Under Secretary of Defense (Acquisition, Technology & Logistics),* Document Acquisition Management Information Retrieval (DAMIR), Selected Acquisition Report (SAR) RCS: DD-A&T(Q&A)823-197, December 31, 1999.

*"DDG 1000," *Office of the Under Secretary of Defense (Acquisition, Technology & Logistics),* Document Acquisition Management Information Retrieval (DAMIR), Selected Acquisition Report (SAR) RCS: DD-A&T(Q&A)823-197, December 31, 2001.

*"DDG 1000," *Office of the Under Secretary of Defense (Acquisition, Technology & Logistics),* Document Acquisition Management Information Retrieval (DAMIR), Selected Acquisition Report (SAR) RCS: DD-A&T(Q&A)823-197, December 31, 2002.

*"DDG 1000," *Office of the Under Secretary of Defense (Acquisition, Technology & Logistics),* Document Acquisition Management Information Retrieval (DAMIR), Selected Acquisition Report (SAR) RCS: DD-A&T(Q&A)823-197, December 31, 2003.

*"DDG 1000," *Office of the Under Secretary of Defense (Acquisition, Technology & Logistics),* Document Acquisition Management Information Retrieval (DAMIR), Selected Acquisition Report (SAR) RCS: DD-A&T(Q&A)823-197, December 31, 2004.

*"DDG 1000," *Office of the Under Secretary of Defense (Acquisition, Technology & Logistics),* Document Acquisition Management Information Retrieval (DAMIR), Selected Acquisition Report (SAR) RCS: DD-A&T(Q&A)823-197, December 31, 2005.

*"DDG 1000," *Office of the Under Secretary of Defense (Acquisition, Technology & Logistics),* Document Acquisition Management Information Retrieval (DAMIR), Selected Acquisition Report (SAR) RCS: DD-A&T(Q&A)823-197, December 31, 2006.

*"DDG 1000," *Office of the Under Secretary of Defense (Acquisition, Technology & Logistics),* Document Acquisition Management Information Retrieval (DAMIR), Selected Acquisition Report (SAR) RCS: DD-A&T(Q&A)823-197, December 31, 2007.

*"DDG 1000," *Office of the Under Secretary of Defense (Acquisition, Technology & Logistics),* Document Acquisition Management Information Retrieval (DAMIR), Selected Acquisition Report (SAR) RCS: DD-A&T(Q&A)823-197, December 31, 2009.

"Department of the Navy Fiscal Year (FY) 2010 Budget Estimates; Justification of Estimates, Shipbuilding and Conversion," *US Navy,* May 2009.

"Department of the Navy Fiscal Year (FY) 2005 Budget Estimates; Justification of Estimates, Shipbuilding and Conversion," *US Navy,* February 2004.

"Department of the Navy Fiscal Year (FY) 2006/FY 2007 Budget Estimates; Justification of Estimates, Shipbuilding and Conversion," *US Navy,* February 2005.

"Department of the Navy Fiscal Year (FY) 2007 Budget Estimates; Justification of Estimates, Shipbuilding and Conversion," *US Navy,* February 2006.

"Department of the Navy Fiscal Year (FY) 2008/2009 Budget Estimates; Justification of Estimates, Shipbuilding and Conversion," *US Navy,* February 2007.

"Department of the Navy Fiscal Year (FY) 2009 Budget Estimates; Justification of Estimates, Shipbuilding and Conversion," *US Navy,* February 2008.

"Department of the Navy Fiscal Year (FY) 2011 Budget Estimates; Justification of Estimates, Shipbuilding and Conversion," *US Navy,* February 2010.

England, Gordon, "Letter to Edward M. Kennedy," *Deputy Secretary of Defense,* August 18, 2008.

Etter, Delores M., "Memorandum for Under Secretary of Defense (Acquisition, Technology & Logistics); Subject: Implementation Plan for Management Controls to Monitor Major Cost Estimate Differences in the DD(X) Destroyer Program," *US Navy, Assistant Secretary of the Navy (Research, Development and Acquisition),* February 16, 2006.

Francis, Paul L., "Testimony Before the Subcommittee on Seapower and Expeditionary Forces, Committee on Armed Services, House of Representatives: Defense Acquisitions: Zumwalt-Class Destroyer Program Emblematic of Challenges Facing Navy Shipbuilding," *United States Government Accountability Office (GAO),* GAO-08-1061T, July 31, 2008.

Gansler, J. S., "Memorandum for Secretary of the Navy Attn: Acquisition Executive; Subject: Acquisition Decision Memorandum for the Navy's Surface Combatant for the 21st Century (SC-21) Program," *Office of the Under Secretary of Defense (Acquisition and Technology),* January 12, 1998.

Gates, Robert, "DoD News Briefing with Secretary Gates from the Pentagon," *Office of the Secretary of Defense,* April 6, 2009.

Horvath, Joseph J., "DDG 1000 Case Study – Overview: Risk Based Source Selection Concept," *DDG 1000 Program Office,* November 7, 2006.

*Krieg, Kenneth J., "Memorandum for Secretary of the Navy, Subject: DDG 1000 Zumwalt Class Destroyer Acquisition Decision Memorandum (ADM)," *Office of the Under Secretary of Defense (Acquisition, Technology & Logistics),* May 24, 2007.

Labs, Eric J., "Testimony Before the Navy's Surface Combatant Programs Before the Subcommittee on Seapower and Expeditionary Forces Committee on Armed Services U.S. House of Representatives," *Congressional Budget Office,* July 31, 2008.

McCullough, Vice Admiral Barry, and Allison Stiller, "Statement Before the Subcommittee on Seapower and Expeditionary Forces of the House Armed Services Committee on Surface Combatant Requirements and Acquisition Strategies," *US Navy,* July 31, 2008.

O'Rourke, Ronald, "CRS Report for Congress: Navy DDG-51 and DDG-1000 Destroyer Programs: Background and Issues for Congress," *Congressional Research Service,* November 23, 2009.

O'Rourke, Ronald, "CRS Report for Congress: Navy DDG-51 and DDG-1000 Destroyer Programs: Background and Issues for Congress," *Congressional Research Service,* June 14, 2010.

O'Rourke, Ron, "Statement of Ronald O'Rourke Specialist in Naval Affairs Congressional Research Before the House Armed Services Committee Subcommittee on Seapower and Expeditionary Forces Hearing on Surface Combatant Warfighting Requirements and Acquisition Strategy," *Congressional Research Service,* July 31, 2008.

"Report to the Subcommittee on Seapower, Committee on Armed Services, U.S. Senate: Cost to Deliver Zumwalt-Class Destroyers Likely to Exceed Budget," *United States Government Accountability Office (GAO),* GAO-08-804, July 2008.

"Report to the Subcommittee on Seapower, Committee on Armed Services, U.S. Senate: Defense Acquisitions: Zumwalt-Class Destroyer Program Emblematic of Challenges Facing Navy Shipbuilding," *United States Government Accountability Office (GAO),* GAO-08-1061T, July 31, 2008.

*Ross, Christopher M., "DDG 1000/Zumwalt Program," *Defense Contract Management Agency,* April 14, 2010.

*Spruill, Dr. Nancy L., "Memorandum for Under Secretary of Defense, USD(AT&L), Subject: DDG-1000 Acquisition Strategy," *OUSD(AT&L)/ARA,* February 1, 2008.

*Syring, Capt. Jim, "Common Link Interface Processor (CLIP) Discussion Presentation to RADM Vic Guillory OPNAV N86," *US Navy, DDG 1000 Program Office,* September 12, 2007.

*Syring, Capt. Jim, "DDG 1000 Program Review Presentation to RADM Barry McCullough OPNAV N8F," *US Navy, DDG 1000 Program Office,* May 17, 2007.

*Syring, Capt. Jim, "Schedule and Cost Review: DDG 1000 Class Destroyer Program," *U.S. Navy, DDG 1000 Program Office,* July 10, 2006.

"The U.S. Navy's Destroyer Acquisition Plan: Examining Options for Acquiring DDG-1000 and DDG-51 Destroyers to Meet Maritime Capability Requirements," *The University of Tennessee, National Defense Business Institute,* 2009.

Weiner, Charles S., "DCMA Raytheon IDS (DDG-1000 ZUMWALT) Quarterly Program Support Team (PST) Predictive Analysis Report," *Defense Contract Management Agency, Space & Missile Systems Division,* April 28, 2009.

Weiner, Charles S., "DCMA Raytheon IDS (DDG-1000 ZUMWALT) Quarterly Program Support Team (PST) Report," *Defense Contract Management Agency, Space & Missile Systems Division,* January 25, 2010.

Winter, Donald C., "Letter to Carl Levin, Chairman, Committee on Armed Services, United States Senate," *US Navy, Secretary of the Navy,* August 18, 2008.

Young, Jr., John J., "Letter to Gene Taylor, Chairman, Subcommittee on Seapower and Expeditionary Forces, Committee on Armed Forces, US House of Representatives," *Office of the Under Secretary of Defense (Acquisition, Technology, and Logistics),* July 2, 2008.

*Young, Jr., John J., "Memorandum for the Record: DDG 1000 Program Way Ahead," *Office of the Under Secretary of Defense (Acquisition, Technology & Logistics),* January 26, 2009.

Data Sources for Excalibur

Note: An asterisk indicates a key source used in the Root Cause Analysis.

Bertuca, Tony, "Army Using HTS, Precision Munitions, Biometrics to 'Win Hearts and Minds,'" *Inside the Army,* April 5, 2010.

*Bolton, Jr., Claude M., "Memorandum for Program Executive Officer, Ammunition: Acquisition Decision Memorandum—Milestone (MS) C Decision for Excalibur XM982 Block Ia-1," Department of the Army, *Office of the Assistant Secretary of the Army (Acquisition, Logistics and Technology),* May 23, 2005.

*Bolton, Jr., Claude M., "Memorandum for Program Executive Officer, Ammunition: Acquisition Decision Memorandum for Low-Rate Initial Production (LRIP) of Excalibur XM982 Block Ia-1," *Department of the Army, Office of the Assistant Secretary of the Army (Acquisition, Logistics and Technology),* March 26, 2007.

*Bolton, Jr., Claude M., "Memorandum for Program Executive Officer, Ammunition: Acquisition Decision Memorandum for Low-Rate Initial Production (LRIP) of Increment 1a-2 Excalibur XM982," *Department of the Army, Office of the Assistant Secretary of the Army (Acquisition, Logistics and Technology),* July 31, 2007.

Brannen, Kate, "Army Moving Quickly on Precision Fire Capability for Afghanistan," *Inside the Army,* May 25, 2009.

Brannen, Kate, "Army Weighing Accuracy, Cost and Design Maturity in Excalibur Ib," *Inside the Army,* November 16, 2009.

"Committee Staff Procurement Backup Book: Fiscal Year (FY) 2008/2009 Budget Estimates: Procurement of Ammunition, Army Appropriation," *Department of the Army, Procurement Programs,* February 2007.

"Committee Staff Procurement Backup Book: Fiscal Year (FY) 2009 Budget Estimates: Procurement of Ammunition, Army Appropriation," *Department of the Army, Procurement Programs,* February 2008.

"Committee Staff Procurement Backup Book: Fiscal Year (FY) 2010 Budget Estimates: Procurement of Ammunition, Army Appropriation," *Department of the Army, Procurement Programs,* May 2009.

"Committee Staff Procurement Backup Book: Fiscal Year (FY) 2011 Budget Estimates: Procurement of Ammunition, Army Appropriation," *Department of the Army, Procurement Programs,* February 2010.

"Critical Intelligence," *Inside the Pentagon,* April 9, 2009.

DiMascio, Jen, "Army Will Purchase Fewer Excalibur Projectiles, Speed Production," *Inside the Army,* April 26, 2004.

"Excalibur," *Office of the Under Secretary of Defense (Acquisition, Technology & Logistics),* Document Acquisition Management Information Retrieval (DAMIR), Acquisition Program Baseline (APB), Approval Date: October 20, 2004.

"Excalibur," *Office of the Under Secretary of Defense (Acquisition, Technology & Logistics),* Document Acquisition Management Information Retrieval (DAMIR), Acquisition Program Baseline (APB), Approval Date: July 27, 2007.

"Excalibur," *Office of the Under Secretary of Defense (Acquisition, Technology & Logistics),* Document Acquisition Management Information Retrieval (DAMIR), SAR Baseline, January 29, 2003.

"Excalibur," *Office of the Under Secretary of Defense (Acquisition, Technology & Logistics),* Document Acquisition Management Information Retrieval (DAMIR), SAR Baseline, January 28, 2008.

*"Excalibur," *Office of the Under Secretary of Defense (Acquisition, Technology & Logistics),* Document Acquisition Management Information Retrieval (DAMIR), Selected Acquisition Report (SAR), December 31, 2002.

*"Excalibur," *Office of the Under Secretary of Defense (Acquisition, Technology & Logistics),* Document Acquisition Management Information Retrieval (DAMIR), Selected Acquisition Report (SAR), December 31, 2003.

*"Excalibur," *Office of the Under Secretary of Defense (Acquisition, Technology & Logistics),* Document Acquisition Management Information Retrieval (DAMIR), Selected Acquisition Report (SAR), December 31, 2004.

*"Excalibur," *Office of the Under Secretary of Defense (Acquisition, Technology & Logistics),* Document Acquisition Management Information Retrieval (DAMIR), Selected Acquisition Report (SAR), December 31, 2005.

*"Excalibur," *Office of the Under Secretary of Defense (Acquisition, Technology & Logistics),* Document Acquisition Management Information Retrieval (DAMIR), Selected Acquisition Report (SAR), December 31, 2006.

*"Excalibur," *Office of the Under Secretary of Defense (Acquisition, Technology & Logistics),* Document Acquisition Management Information Retrieval (DAMIR), Selected Acquisition Report (SAR), September 30, 2007.

*"Excalibur," *Office of the Under Secretary of Defense (Acquisition, Technology & Logistics),* Document Acquisition Management Information Retrieval (DAMIR), Selected Acquisition Report (SAR), December 31, 2007.

*"Excalibur," *Office of the Under Secretary of Defense (Acquisition, Technology & Logistics),* Document Acquisition Management Information Retrieval (DAMIR), Selected Acquisition Report (SAR), December 31, 2009.

Feiler, Jeremy, "Army Requests Money to Develop, Buy Excalibur Precision Round," *Inside the Army,* February 5, 2004.

John, Libby, "Third Variant of Excalibur Munition Delayed for Up to a Year," *Inside the Army,* May 22, 2006.

Maffei, Glenn, "Army Officially Commits to Early Fielding of Excalibur Munition," *Inside the Army,* April 11, 2005.

"Making Headlines This Week," *Inside the Air Force,* December 10, 2004.

Malenic, Marina, "Army Study Recommends Equipping Infantry with Precision Mortars," *Inside the Army,* November 19, 2007.

Malenic, Marina, "Army Wants More Affordable 155 Millimeter Precision Munition," *Inside the Army,* December 3, 2007.

*McHugh, John M., "Letters of Notification to Congress," *US Army, Secretary of the Army,* August 20, 2010.

"Memorandum for: Project Manager, Combat Ammo Systems, SFAE-AMO-CASEX LTC Joe Minus; Excalibur Program Monthly Report for the period of 14 January 2008 – 17 February 2008," *Defense Contract Management Agency, DCMA Missile Operations, Raytheon – Tucson,* February 20, 2008.

"Memorandum for: Project Manager, Combat Ammo Systems, SFAE-AMO-CASEX, LTC Joe Minus; Excalibur Program Monthly Report for the period of 17 February 2008 – March 20," *Defense Contract Management Agency, DCMA Missile Operations, Raytheon – Tucson,* March 20, 2008.

"Memorandum for: Project Manager, Combat Ammo Systems, SFAE-AMO-CASEX, LTC Joe Minus; Excalibur Program Monthly Report for the period of March 20 – April 15, 2008," *Defense Contract Management Agency, DCMA Missile Operations, Raytheon – Tucson,* April 15, 2008.

"Memorandum for: Project Manager, Combat Ammo Systems, SFAE-AMO-CASEX LTC Joe Minus; Excalibur Program Monthly Report for the period of April 15 – May 20, 2008," *Defense Contract Management Agency, DCMA Missile Operations, Raytheon – Tucson,* May 2008.

"Memorandum for: Project Manager, Combat Ammo Systems, SFAE-AMO-CASEX, LTC Joe Minus; Excalibur Program Monthly Report for the period of 20 May – 13 June," *Defense Contract Management Agency, DCMA Missile Operations, Raytheon – Tucson,* June 13, 2008.

"Memorandum for: Project Manager, Combat Ammo Systems, SFAE-AMO-CASEX LTC Joe Minus; Excalibur Program Monthly Report for the period of 13 June – 10 July," *Defense Contract Management Agency, DCMA Missile Operations, Raytheon – Tucson,* July 2008.

"Memorandum for: Project Manager, Combat Ammo Systems, SFAE-AMO-CASEX LTC Joe Minus; Excalibur Program Monthly Report for the period of 15 July – 13 August, 2008," *Defense Contract Management Agency, DCMA Missile Operations, Raytheon – Tucson,* August 13, 2008.

"Memorandum for: Project Manager, Combat Ammo Systems, SFAE-AMO-CASEX LTC Joe Minus; Excalibur Program Monthly Report for the period of 13 Aug – 17 Sept 08," *Defense Contract Management Agency, DCMA Missile Operations, Raytheon – Tucson,* September 17, 2008.

"Memorandum for: Project Manager, Combat Ammo Systems, SFAE-AMO-CASEX LTC Joe Minus; Excalibur Program Monthly Report for the period of 17 Sept – 8 Oct 2008," *Defense Contract Management Agency, DCMA Missile Operations, Raytheon – Tucson,* October 8, 2008.

"Memorandum for: Project Manager, Combat Ammo Systems, SFAE-AMO-CASEX LTC Joe Minus; Excalibur Program Monthly Report for the period of 8 Oct –18 Nov 08," *Defense Contract Management Agency, DCMA Missile Operations, Raytheon – Tucson,* November 18, 2008.

"Memorandum for: Project Manager, Combat Ammo Systems, SFAE-AMO-CASEX, LTC Joe Minus; Excalibur Program Status Report 18 Nov – 11 Dec 08," *Defense Contract Management Agency, DCMA Missile Operations, Raytheon – Tucson,* December 11, 2008.

"Memorandum for: Project Manager, Combat Ammo Systems, SFAE-AMO-CASEX, LTC Joe Minus; Excalibur Program Status Report 11 Dec – 20 Jan 09," *Defense Contract Management Agency, DCMA Missile Operations, Raytheon – Tucson,* January 21, 2009.

"Memorandum for: Project Manager, Combat Ammo Systems, SFAE-AMO-CASEX, LTC Joe Minus; Excalibur Program Status Report 20 Jan – Feb 6 09,"

Defense Contract Management Agency, DCMA Missile Operations, Raytheon – Tucson, February 12, 2009.

"Memorandum for: Project Manager, Combat Ammo Systems, SFAE-AMO-CASEX, LTC Joe Minus; Excalibur Program Status Report 6 Feb – 8 Mar 09," *Defense Contract Management Agency, DCMA Missile Operations, Raytheon – Tucson,* March 13, 2009.

"Memorandum for: Project Manager, Combat Ammo Systems, SFAE-AMO-CASEX, LTC Joe Minus; DCMA Excalibur Program Status Report March 09," *Defense Contract Management Agency, DCMA Missile Operations, Raytheon – Tucson,* April 15, 2009.

"Memorandum for: Project Manager, Combat Ammo Systems, SFAE-AMO-CASEX, LTC Joe Minus; Excalibur Program Status Report April 15 – 12 May 09," *Defense Contract Management Agency, DCMA Missile Operations, Raytheon – Tucson,* May 15, 2009.

"Memorandum for: Project Manager, Combat Ammo Systems, SFAE-AMO-CASEX, LTC Joe Minus; Excalibur Program Status Report 15 May – 15 June 09," *Defense Contract Management Agency, DCMA Missile Operations, Raytheon – Tucson,* June 17, 2009.

"Memorandum for: Project Manager, Combat Ammo Systems, SFAE-AMO-CASEX, LTC Joe Minus; Excalibur Program Status Report 15 June – 15 July 09," *Defense Contract Management Agency, DCMA Missile Operations, Raytheon – Tucson,* July 17, 2009.

"Memorandum for: Project Manager, Combat Ammo Systems, SFAE-AMO-CASEX, LTC Joe Minus; Excalibur Program Status Report 16 July – 11 Aug 2009," *Defense Contract Management Agency, DCMA Missile Operations, Raytheon – Tucson,* August 12, 2009.

"Memorandum for: Project Manager, Combat Ammo Systems, SFAE-AMO-CASEX, LTC Milner; Excalibur Program Monthly Report for the period of 11 Aug – Sept. 11, 2009," *Defense Contract Management Agency, DCMA Missile Operations, Raytheon – Tucson,* September 14, 2009.

"Memorandum for: Program Manager, Combat Ammo Systems, SFAE-AMO-CASEX, LTC Milner; Excalibur Program Status Report Sept. 11 – Oct 15, 2009," *Defense Contract Management Agency, DCMA Missile Operations, Raytheon – Tucson,* October 19, 2009.

"Memorandum for: Program Manager, Combat Ammo Systems, SFAE-AMO-CASEX, LTC Milner; Excalibur Program Status Report Oct 15 – Nov 12," *Defense Contract Management Agency, DCMA Missile Operations, Raytheon – Tucson,* November 13, 2009.

"Memorandum for: Program Manager, Combat Ammo Systems, SFAE-AMO-CASEX, LTC Milner; Excalibur Program Status Report Nov 12 – Dec 11," *Defense Contract Management Agency, DCMA Missile Operations, Raytheon – Tucson,* December 15, 2009.

"Memorandum for: Program Manager, Combat Ammo Systems, SFAE-AMO-CASEX, LTC Milner; Excalibur Program Status Report Dec 11, 2009 – Jan 12, 2010," *Defense Contract Management Agency, DCMA Missile Operations, Raytheon – Tucson,* January 15, 2010.

"Memorandum for: Program Manager, Combat Ammo Systems, SFAE-AMO-CASEX, LTC Milner; Excalibur Program Status Report Jan 12 – Feb 15, 2010," *Defense Contract Management Agency, DCMA Missile Operations, Raytheon – Tucson,* February 18, 2010.

"Memorandum for: Program Manager, Combat Ammo Systems, SFAE-AMO-CASEX, LTC Milner; Excalibur Program Status Report Feb 15 – March 15, 2010," *Defense Contract Management Agency, DCMA Missile Operations, Raytheon – Tucson,* March 18, 2010.

"Memorandum for: Program Manager, Combat Ammo Systems, SFAE-AMO-CASEX, LTC Milner; Excalibur Program Status Report March 15 – April 15, 2010," *Defense Contract Management Agency, DCMA Missile Operations, Raytheon – Tucson,* April 21, 2010.

"Memorandum for: Program Manager, Combat Ammo Systems, SFAE-AMO-CASEX, LTC Milner; Excalibur Program Status Report April 15 – May 14, 2010," *Defense Contract Management Agency, DCMA Missile Operations, Raytheon – Tucson,* May 18, 2010.

"Memorandum for: Program Manager, Combat Ammo Systems, SFAE-AMO-CASEX, LTC Milner; Excalibur Program Status Report May 14 – June 15, 2010," *Defense Contract Management Agency, DCMA Missile Operations, Raytheon – Tucson,* June 18, 2010.

"Memorandum for: Program Manager, Combat Ammo Systems, SFAE-AMO-CASEX, LTC Milner; Excalibur Program Status Report June 15 – July 2010," Defense Contract Management Agency, *DCMA Missile Operations, Raytheon – Tucson,* July 16, 2010.

"Memorandum for: Program Manager, Combat Ammo Systems, SFAE-AMO-CASEX, LTC Milner; Excalibur Program Status Report July 16 – August 17, 2010," Defense Contract Management Agency, *DCMA Missile Operations, Raytheon – Tucson,* August 17, 2010.

"Memorandum for: Program Manager, Combat Ammo Systems, SFAE-AMO-CASEX, LTC Milner; Excalibur Monthly Program Analysis Report for 1a Production and 1b System technical Design and Performance," *Defense Contract*

Management Agency, DCMA Missile Operations, Raytheon – Tucson, September 20, 2010.

*Milner, LTC Michael, "Excalibur XM982: Brief to Nunn-McCurdy IPT," *US Army, Product Manager for Excalibur* (undated).

*Popps, Dean G., "Memorandum for Program Executive Officer, Ammunition: Excalibur: Acquisition Decision Memorandum," *Department of the Army, Office of the Assistant Secretary of the Army (Acquisition, Logistics and Technology)*, January 8, 2009.

"Report to Congressional Committees: Defense Acquisitions: Assessments of Major Weapon Programs," *United States Government Accountability Office (GAO)*, GAO-04-248, March 2004.

"Report to Congressional Committees: Defense Acquisitions: Assessments of Major Weapon Programs," *United States Government Accountability Office (GAO)*, GAO-05-301, March 2005.

"Report to Congressional Committees: Defense Acquisitions: Assessments of Major Weapon Programs," *United States Government Accountability Office (GAO)*, GAO-06-391, March 2006.

"Report to Congressional Committees: Defense Acquisitions: Assessments of Major Weapon Programs," *United States Government Accountability Office (GAO)*, GAO-07-406SP, March 2007.

"Report to Congressional Committees: Defense Acquisitions: Assessments of Major Weapon Programs," *United States Government Accountability Office (GAO)*, GAO-08-467SP, March 2008.

"Report to Congressional Committees: Defense Acquisitions: Assessments of Major Weapon Programs," *United States Government Accountability Office (GAO)*, GAO-09-326SP, March 2009.

"Report to Congressional Committees: Defense Acquisitions: Assessments of Major Weapon Programs," *United States Government Accountability Office (GAO)*, GAO-10-388SP, March 2010.

Roque, Ashley, "Army Efforts to Rush Excalibur to the Field Slowed by One Year," *Inside the Army*, December 4, 2006.

Roque, Ashley, "Changes to Excalibur in the Works After Flight Testing Goes Awry," *Inside the Army*, August 7, 2006.

Scully, Megan, "Acquisition Category Upgraded for Programs Sped by Crusader Kill," *Inside the Army*, June 17, 2002.

Siegelbaum, Debbie, "Excalibur 'Critical' Nunn-McCurdy Program Deviation Report Delivered," *Inside the Army,* July 26, 2010.

Sprenger, Sebastian, "After NLOS-LS, Army Ponders New Fire-Support Options for Infantry BCTs," *Inside the Army,* June 7, 2010.

Sprenger, Sebastian, "Army Weighs New Hedge Against Cost Growth in Precision-Fires Arena," *Inside the Army,* July 12, 2010.

Sprenger, Sebastian, "Reasons for Low Wartime Excalibur Use Hard to Pin Down, Officials Say," *Inside the Army,* July 5, 2010.

Tellez, Patty, "Excalibur M982 Artillery Round," *DCMA Raytheon Tucson,* September 30, 2010.

*Turner, Jr., COL J. Scott, "Memorandum Thru: Program Executive Officer Ammunition (SFAE-AMO); For Assistant Secretary of the Army (Acquisition, Logistics and Technology (SAAL-ZA): Excalibur Program Deviation Report," *US Army, Project Manager, Combat Ammunition Systems,* July 6, 2010.

*Westphal, Joseph W., "Memorandum for Program Executive Officer, Ammunition: Excalibur Program Acquisition Decision Memorandum," *US Army, Under Secretary of the Army,* May 12, 2010.

*Wiltz, James, "Excalibur Nunn-McCurdy Preliminary Information," *Office of the Under Secretary of Defense (Acquisition, Technology & Logistics),* August 3, 2010.

*Wiltz, James, "Excalibur Nunn-McCurdy Kick-off Meeting," *Office of the Under Secretary of Defense (Acquisition, Technology & Logistics),* August 20, 2010.

Winograd, Erin Q., "Top Army Acquisition Official Decries Lack of Program Stability," *Inside the Army,* June 26, 2000.

Winograd, Erin Q., "U.S. and Sweden Agree to Joint Development Program for Excalibur," *Inside the Army,* May 6, 2002.

Data Sources for the Joint Strike Fighter

Note: An asterisk indicates a key source used in the Root Cause Analysis.

"2007 JSF Commonality Database," *Joint Strike Fighter Program Office,* September 6, 2007.

"2008 JSF Commonality Database," *Joint Strike Fighter Program Office,* July 30, 2008.

"240-4 JSF Commonality Database," *Joint Strike Fighter Program Office,* December 1, 2004.

"240-4.2 JSF Commonality Database," *Joint Strike Fighter Program Office,* September 21, 2005.

"240-4.5 JSF Commonality Database," *Joint Strike Fighter Program Office,* November 14, 2006.

*Aldridge, Jr., E. C., "Memorandum for the Secretary of the Air Force and Secretary of the Navy: Subject: Joint Strike Fighter (JSF) Program Milestone B Acquisition Decision Memorandum," *Office of the Under Secretary of Defense (Acquisition, Technology and Logistics),* October 26, 2001.

*Aldridge, Jr., E. C., "Memorandum for the Secretary of the Air Force and Secretary of the Navy; Vice Chairman, Joint Chiefs of Staff: Subject: Joint Strike Fighter (JSF) Program Milestone B Acquisition Decision Memorandum," *Office of the Under Secretary of Defense (Acquisition, Technology and Logistics),* July 26, 2002.

"Analysis of the Growth in Funding for Operations in Iraq, Afghanistan, and Elsewhere in the War on Terrorism," *Congressional Budget Office,* February 11, 2008.

"AS CDR 3 Summary Assessment," *Lockheed Martin Corporation,* June 18–22, 2007.

"Assessment Roadmap Period 11," *Joint Strike Fighter Program,* date unknown.

*Baker, David A., "F-35 Lightning II: PARCA Review Program Management," *Lockheed Martin*, April 23, 2010.

Balderson, William, "Memorandum for the Record: Joint Strike Fighter (JSF) Acquisition Program Baseline Agreement (APBA)," *US Navy, Deputy Assistant Secretary of the Navy (Air Programs)*, February 26, 2007.

Birkler, John L., John C. Graser, Mark V. Arena, Cynthia R. Cook, Gordon T. Lee, Mark A. Lorell, Giles K. Smith, Fred S. Timson, Obaid Younossi, and Jonathan Gary Grossman, "Assessing Competitive Strategies for the Joint Strike Fighter: Opportunities and Options," *RAND Corporation*, MG-1362-OSD/JSF, 2001.

Bolkcom, Christopher, "Joint Strike Fighter (JSF) Program: Background, Status, and Issues," *Congressional Research Service*, February 15, 2002.

Butler, Amy, "Joint Strike Fighter; Lockheed CEO: Further F-35 Stovl 'Rephasing' Possible," *Aviation Week & Space Technology*, September 13, 2010, p. 29.

*Butler, Amy, "Lockheed Martin Reprimanded for Poor Auditing at JSF Plant," *Aviation Week & Space Technology*, October 11, 2010, p. 26.

*Carter, Ashton B., "Memorandum for Secretary of the Navy and Secretary of the Air Force; Subject: F-35 Lightning II Joint Strike Fighter (JSF) Program Restructure Acquisition Decision Memorandum (ADM)," *Office of the Under Secretary of Defense (Acquisition, Technology and Logistics)*, February 24, 2010.

*Carter, Ashton B., "Testimony of Ashton B. Carter Under Secretary of Defense (Acquisition, Technology & Logistics) Before the United States House Committee on Armed Services Air and Land Forces Subcommittee and Seapower and Expeditionary Forces Subcommittee," *US House of Representatives*, March 24, 2010.

Cohen, Randall, "Letter to Lockheed Martin on Special Contract Provision H-3 for fifth award fee evaluation period," *Joint Strike Fighter Program*, September 30, 2003.

Cohen, Randall, "Update to Letter to Lockheed Martin on Special Contract Provision H-3 for fifth award fee evaluation period," *Joint Strike Fighter Program*, October 31, 2003.

Cohen, Randall S., "Update to Letter to Lockheed Martin on Special Contract Provision H-3 for sixth award fee evaluation period," *Joint Strike Fighter Program*, July 23, 2004.

*Davis, Brig Gen Charles R., "Memorandum for the Under Secretary of Defense (Acquisition, Technology and Logistics) via Assistant Secretary of the Navy (Research, Development and Acquisition); Subject: Joint Strike Fighter (JSF) Program Deviation Report," *Joint Strike Fighter Program Office*, November 21, 2006.

*Davis, Brig Gen Charles R., "Memorandum for the Under Secretary of Defense (Acquisition, Technology and Logistics) via Assistant Secretary of the Navy (Research, Development and Acquisition); Subject: Joint Strike Fighter (JSF) Program Deviation Report Dated 21 Nov 2006 Addendum," *Joint Strike Fighter Program Office,* February 2, 2007.

"Department of Defense Appropriations Bill, 2008, Report of the Committee on Appropriations," *US House of Representatives,* July 30, 2007.

*Donley, Michael B., "Letter to the Honorable Carl Levin, Committee on Armed Services Regarding Nunn-McCurdy Breach," *US Air Force, Office of the Secretary of the Air Force,* March 25, 2010.

*Drezner, Jeffrey A., "The Nature and Role of Prototyping in Weapon System Development," *RAND Corporation,* R-4161-ACQ, 1992.

*Etter, Delores M., "Letter to Dan Crowley, Lockheed Martin Corporation on Duplicate Billing Charges," *US Navy, Assistant Secretary of the Navy (Research, Development and Acquisition),* August 10, 2007.

"F-35," *Office of the Under Secretary of Defense (Acquisition, Technology & Logistics),* Document Acquisition Management Information Retrieval (DAMIR), Acquisition Program Baseline (APB), Approval Date: November 15, 1996.

"F-35," *Office of the Under Secretary of Defense (Acquisition, Technology & Logistics),* Document Acquisition Management Information Retrieval (DAMIR), Acquisition Program Baseline (APB), Approval Date: October 26, 2001.

"F-35," *Office of the Under Secretary of Defense (Acquisition, Technology & Logistics),* Document Acquisition Management Information Retrieval (DAMIR), Acquisition Program Baseline (APB), Approval Date: March 17, 2004.

"F-35," *Office of the Under Secretary of Defense (Acquisition, Technology & Logistics),* Document Acquisition Management Information Retrieval (DAMIR), Acquisition Program Baseline (APB), Approval Date: March 30, 2007.

"F-35," *Office of the Under Secretary of Defense (Acquisition, Technology & Logistics),* Document Acquisition Management Information Retrieval (DAMIR), SAR Baseline, November 16, 1996.

"F-35," *Office of the Under Secretary of Defense (Acquisition, Technology & Logistics),* Document Acquisition Management Information Retrieval (DAMIR), SAR Baseline, January 14, 2003.

*"F-35," *Office of the Under Secretary of Defense (Acquisition, Technology & Logistics),* Document Acquisition Management Information Retrieval (DAMIR), Selected Acquisition Report (SAR) RCS: DD-A&T(Q&A)823-197, December 31, 1997.

*"F-35," *Office of the Under Secretary of Defense (Acquisition, Technology & Logistics)*, Document Acquisition Management Information Retrieval (DAMIR), Selected Acquisition Report (SAR) RCS: DD-A&T(Q&A)823-197, December 31, 1998.

*"F-35," *Office of the Under Secretary of Defense (Acquisition, Technology & Logistics)*, Document Acquisition Management Information Retrieval (DAMIR), Selected Acquisition Report (SAR) RCS: DD-A&T(Q&A)823-197, December 31, 1999.

*"F-35," *Office of the Under Secretary of Defense (Acquisition, Technology & Logistics)*, Document Acquisition Management Information Retrieval (DAMIR), Selected Acquisition Report (SAR) RCS: DD-A&T(Q&A)823-197, September 30, 2001.

*"F-35," *Office of the Under Secretary of Defense (Acquisition, Technology & Logistics)*, Document Acquisition Management Information Retrieval (DAMIR), Selected Acquisition Report (SAR) RCS: DD-A&T(Q&A)823-197, December 31, 2001.

*"F-35," *Office of the Under Secretary of Defense (Acquisition, Technology & Logistics)*, Document Acquisition Management Information Retrieval (DAMIR), Selected Acquisition Report (SAR) RCS: DD-A&T(Q&A)823-197, December 31, 2002.

*"F-35," *Office of the Under Secretary of Defense (Acquisition, Technology & Logistics)*, Document Acquisition Management Information Retrieval (DAMIR), Selected Acquisition Report (SAR) RCS: DD-A&T(Q&A)823-197, December 31, 2003.

*"F-35," *Office of the Under Secretary of Defense (Acquisition, Technology & Logistics)*, Document Acquisition Management Information Retrieval (DAMIR), Selected Acquisition Report (SAR) RCS: DD-A&T(Q&A)823-197, December 31, 2004.

*"F-35," *Office of the Under Secretary of Defense (Acquisition, Technology & Logistics)*, Document Acquisition Management Information Retrieval (DAMIR), Selected Acquisition Report (SAR) RCS: DD-A&T(Q&A)823-197, December 31, 2005.

*"F-35," *Office of the Under Secretary of Defense (Acquisition, Technology & Logistics)*, Document Acquisition Management Information Retrieval (DAMIR), Selected Acquisition Report (SAR) RCS: DD-A&T(Q&A)823-197, December 31, 2006.

*"F-35," *Office of the Under Secretary of Defense (Acquisition, Technology & Logistics)*, Document Acquisition Management Information Retrieval (DAMIR), Selected Acquisition Report (SAR) RCS: DD-A&T(Q&A)823-197, December 31, 2007.

*"F-35," *Office of the Under Secretary of Defense (Acquisition, Technology & Logistics)*, Document Acquisition Management Information Retrieval (DAMIR), Selected Acquisition Report (SAR) RCS: DD-A&T(Q&A)823-197, December 31, 2008.

*"F-35," *Office of the Under Secretary of Defense (Acquisition, Technology & Logistics)*, Document Acquisition Management Information Retrieval (DAMIR), Selected Acquisition Report (SAR) RCS: DD-A&T(Q&A)823-197, December 31, 2009.

*Fox, Christine H., "Testimony of Christine H. Fox, Director, Cost Assessment and Program Evaluation, Office of the Secretary of Defense Before the United States House Committee on Armed Services Air and Land Forces Subcommittee and Seapower and Expeditionary Forces Subcommittee," *US House of Representatives,* March 24, 2010.

"FY05 Congressional Summary," *Joint Strike Fighter Program Office,* date unknown.

"FY06 Congressional Summary," *Joint Strike Fighter Program Office,* date unknown.

"FY07 Congressional Summary," *Joint Strike Fighter Program Office,* October 17, 2006.

"FY08 Congressional Summary," *Joint Strike Fighter Program Office,* date unknown.

"FY09 Congressional Summary," *Joint Strike Fighter Program Office,* date unknown.

*"FY10 Congressional Summary," *Joint Strike Fighter Program Office,* December 16, 2009.

Gertler, Jeremiah, "F-35 Joint Strike Fighter (JSF) Program: Background and Issues for Congress," *Congressional Research Service,* December 22, 2009.

Gertler, Jeremiah, "F-35 Alternate Engine Program: Background and Issues for Congress," *Congressional Research Service,* March 22, 2010.

Gertler, Jeremiah, "F-35 Joint Strike Fighter (JSF) Program: Background and Issues for Congress," *Congressional Research Service,* April 2, 2010.

*Gilmore, Dr. J. Michael, "Testimony of Dr. J. Michael Gilmore, Director, Operational Test and Evaluation, Office of the Secretary of Defense Before the United States House Armed Services Committee Joint Seapower and Expeditionary Forces and Air and Land Forces Subcommittees," *United States House of Representatives,* March 24, 2010.

Graser, Jack, and Rob Leonard, "Project Air Force Briefing: JSF Cost Growth—A History from SAR Data," *RAND Corporation,* March 2009.

Gruetzmacher, John, "Joint Strike Fighter: Interoperable by Design," *Lockheed Martin,* October 22, 2003.

Hartley, Richard K., "Memorandum for AF/ASP, SAF/AQP and SAF/FMBI; Subject: 30 Day F-35 Cost Assessment," *US Air Force, Air Force Cost Analysis Agency,* May 21, 2008.

*Hill, Richard, "F-35 Lightning II: Factory Flow," *Lockheed Martin,* April 19, 2010.

*Hilton, Jon, "F-35 Lightning II: Business Systems Transformation," *Lockheed Martin*, April 20, 2010.

*"Implementing Management for Performance and Related Reforms to Obtain Value in Every Acquisition Act of 2010," *US House of Representatives*, April 23, 2010.

"John Warner National Defense Authorization Act for Fiscal Year 2007 – Conference Report to Accompany H.R. 5122," *US House of Representatives*, September 29, 2006.

"Joint Strike Fighter: Additional Costs and Delays Risk Not Meeting Warfighter Requirements on Time, Highlights" *United States Government Accountability Office (GAO)*, GAO-10-382, March 2009.

"JSF PSFD MOU," *Joint Strike Fighter Program Office*, November 14, 2006.

*"JSF SDD Master Production Schedule Evolution," *Lockheed Martin Aeronautics Company*, April 20, 2010.

Kaminski, Paul G., "Joint Strike Fighter (JSF) Milestone I Acquisition Decision Memorandum (ADM)," *Office of the Under Secretary of Defense (Acquisition and Technology)*, November 15, 1996.

Kenne, Brigadier General Leslie, "The Affordable Solution – JSF," *Joint Strike Fighter Program Office*, date unknown.

*Kinard, Don, "F-35 Lightning II: IMRT Overview," *Lockheed Martin*, April 22, 2010.

*Krieg, Kenneth J., "Memorandum for Assistant Secretary of the Navy for Research, Development and Acquisition; Subject: F-35 (Joint Strike Fighter) Lightning II Acquisition Program Baseline Development Estimate Change 2," *Office of the Under Secretary of Defense (Acquisition, Technology and Logistics)*, May 30, 2007.

*"Long-Term Implications of the Fiscal Year 2010 Defense Budget," *Congressional Budget Office*, January 2010.

"Making Appropriations for the Department of Defense for the Fiscal Year Ending September 30, 2007, and Other Purposes – Conference Report to Accompany H.R. 5631," *US House of Representatives*, September 25, 2006.

*Marshall, Guy, "IMRT Funding Strategy & Schedule for Producibility," Lockheed Martin, April 13, 2010.

McNicol, David L., "Joint Acquisition: Implications from Experience with Fixed-Wing Tactical Aircraft," *Institute for Defense Analyses*, September 2005.

McQuain, Paul M., "F-35 Lightning II Program Executive Summary," *Defense Contract Management Agency Lockheed Martin,* Ft Worth, April 14, 2010.

Melton, Maria V., "Letter to Lockheed Martin on Special Contract Provision H-3 for third award fee evaluation period," *Joint Strike Fighter Program,* September 30, 2002.

Melton, Maria V., "Letter to Lockheed Martin on Special Contract Provision H-3 for fourth award fee evaluation period," *Joint Strike Fighter Program,* March 31, 2003.

Melton, Maria V., "Letter to Lockheed Martin on Special Contract Provision H-3 for sixth award fee evaluation period," *Joint Strike Fighter Program,* March 31, 2004.

"National Defense Authorization Act for Fiscal Year 2008," *Committee on Armed Services, United States Senate,* June 5, 2007.

*"N-M Questions Addressed in IPT 5," *Joint Strike Fighter Program,* April 9, 2010.

Norris, Guy, "P&W Reveals Top Thrust Capabilities in JSF Power Battle," *Aviation Week & Space Technology,* September 6, 2010, p. 30.

Norris, Guy, "Tests of Alternate JSF Engine Show Higher Thrust," *Aviation Week & Space Technology,* August 16, 2010, p. 18.

*Novak, Michael, "Executive Summary Memorandum for Under Secretary of Defense (Acquisition, Technology and Logistics) and Principal Deputy Under Secretary of Defense (Acquisition, Technology and Logistics) through Director, Defense Research and Engineering; Subject: JSF Environmental Issues – Status," *Office of the Under Secretary of Defense (Acquisition, Technology and Logistics)/ Strategic and Tactical Systems,* December 13, 2000.

O'Rourke, Ronald, "F-35 Joint Strike Fighter (JSF) Program: Background and Issues for Congress," *Congressional Research Service,* September 16, 2009.

Ozdemir, Levent, "Analyzing the Multi-national Cooperative Acquisition Aspect of the Joint Strike Fighter (JSF) Program," *Naval Post Graduate School,* December 2009.

*Park, Paul, "F-35 Lightning II: PARCA Review," *Lockheed Martin,* April 23, 2010.

Pozda, Andrew F., "OIPT Read Ahead," *Office of the Under Secretary of Defense (Acquisition, Technology and Logistics),* November 16, 2005.

"Report to Congressional Committees; Defense Acquisitions: Assessments of Selected Major Weapon Programs," *United States Government Accountability Office (GAO),* GAO-06-391, March 2006.

"Report to Congressional Committees: Joint Strike Fighter, Additional Costs and Delays Risk Not Meeting Warfighter Requirements on Time," *United States Government Accountability Office (GAO)*, GAO-10-382, March 2010.

"Report to Congressional Committees; Joint Strike Fighter: DOD Plans to Enter Production Before Testing Demonstrates Acceptable Performance," *United States Government Accountability Office (GAO)*, GAO-06-356, March 2006.

"Report to Congressional Committees: Joint Strike Fighter, Recent Decisions by DOD Add to Program Risks," *United States Government Accountability Office (GAO)*, GAO-08-388, March 2008.

"Report to the Chairman, Subcommittee on Air and Land Forces, Committee on Armed Services, House of Representatives; Tactical Aircraft DOD Needs a Joint and Integrated Investment Strategy," *United States Government Accountability Office (GAO)*, GAO-07-415, April 2007.

"SDD Master Schedule," *Lockheed Martin Aeronautics Company,* November 30, 2001.

*"Statement of Chairman Adam Smith Joint Hearing of the Air and Land Forces and Seapower and Expeditionary Forces Subcommittees Department of the Navy and Air Force Combat Aviation Programs," *US House of Representatives,* March 24, 2010.

*"Statement of Chairman Gene Taylor Joint Hearing of the Air and Land Forces and Seapower and Expeditionary Forces Subcommittees Department of the Navy and Air Force Combat Aviation Programs," *US House of Representatives,* March 24, 2010.

*"Statement of the Honorable Sean J. Stackley, Assistant Secretary of the Navy (Research, Development and Acquisition) and Lieutenant General George J. Trautman III, USMC Deputy Commandant for Aviation and Rear Admiral David L. Philman, USN Director of Air Warfare Before the Seapower and Expeditionary Forces and Air and Land Forces Subcommittees of the House Armed Services Committee on Department of the Navy's Aviation Procurement Program," *US House of Representatives,* March 24, 2010.

Sullivan, Michael, "Testimony Before the Subcommittee on AirLand, Committee on Armed Services, U.S. Senate; Tactical Aircraft Recapitalization Goals Are Not Supported by Knowledge-Based F-22A and JSF Business Cases," *United States Government Accountability Office (GAO)*, GAO-06-523T, March 28, 2006.

Sullivan, Michael, "Testimony Before the Subcommittee on Air and Land Forces, Committee on Armed Services, House of Representatives; Joint Strike Fighter Strong Risk Management Essential as Program Enters Most Challenging Phase," *United States Government Accountability Office (GAO)*, GAO-09-711T, May 20, 2009.

*Sullivan, Michael, "Testimony Before the Committee on Armed Services, U.S. Senate; Joint Strike Fighter Significant Challenges Remain as DOD Restructures Program," *United States Government Accountability Office (GAO)*, GAO-10-520T, March 11, 2010.

*Sullivan, Michael, "United States Government Accountability Office Testimony Before the Subcommittees on Air and Land Forces and Seapower and Expeditionary Forces, Committee on Armed Services, House of Representatives; Joint Strike Fighter Significant Challenges and Decisions Ahead," *United States Government Accountability Office (GAO)*, GAO-10-478T, March 24, 2010.

Sweetman, Bill, "Marines Ready, But Is JSF?," *Defense Technology International*, October 1, 2010, p. 32.

Thoden, Brent, and Jim Ruocco, "Introduction & CDR Process AS CDR 3," *Lockheed Martin Corporation*, June 19, 2007.

*Van Buren, David M., "Department of the Air Force Presentation to the House Armed Services Committee Air and Land Forces Subcommittee and Seapower and Expeditionary Forces Subcommittee," *US House of Representatives*, March 24, 2010.

Warwick, Graham, "Flight Tests of Next F-35 Mission-System Block Underway," *Aviation Week & Space Technology*, June 7, 2010, p. 26.

Warwick, Graham, "Growth in Stovl F-35 Flying Lags New Test Plan," *Aviation Week & Space Technology*, July 19, 2010, p. 70.

Warwick, Graham, "F-35 Flights Suspended by Fuel Pump, Inlet Door Issues," *Aerospace Daily & Defense Report*, October 4, 2010, p. 12.

Warwick, Graham, "Pratt Completes F135 Overtemp and Durability Tests," *Aerospace Daily & Defense Report*, October 19, 2010, p. 01.

Warwick, Graham, "Study Will Assess Alternate Ejection Seat for F-35," *Aerospace Daily & Defense Report*, September 20, 2010, p. 6.

Weisgerber, Marcus, "DOD Looks to Integrate Intelligence Expertise into Acquisition Process," *Inside the Air Force*, April 30, 2010.

Data Sources for the Navy Enterprise Resource Planning Program

Note: An asterisk indicates a key source used in the Root Cause Analysis.

*"Acquisition Program Baseline Agreement: Navy ERP," *US Navy, Navy Enterprise Resource Planning Program,* September 18, 2007.

*Carter, Jennifer, "FY10 Re-certification Briefing to Dr. Nancy L. Spruill," *US Navy, Navy Enterprise Resource Planning (ERP) Program Office,* July 14, 2010.

*Carter, Jennifer, "NAVSEA Phased Deployment Decision & FY11 Re-certification Briefing to Investment Review Board," *US Navy, Navy Enterprise Resource Planning (ERP) Program Office,* August 18, 2010.

*Carter, Jennifer, "Navy ERP Program Overview," *US Navy, Navy Enterprise Resource Planning (ERP) Program Office,* December 2, 2010.

*"Changes from Milestone C," *US Navy, Navy Enterprise Resource Planning (ERP) Program Office,* December 2, 2010.

*Commons, Gladys J., and Sean Stackley, "Memorandum for Distribution; Subject: Department of the Navy Service Cost Positions," *Department of the Navy, Office of the Secretary,* January 7, 2010.

*"Financial Extension (FEX) Acquisition Approach: Expanded Acquisition Approach Considerations – Discussions w/ OSD," *US Navy, Navy Enterprise Resource Planning (ERP) Program Office,* September 14, 2010.

*McGrath, Elizabeth A., "Memorandum for Assistant Secretary of the Navy (Research, Development and Acquisition); Subject: Navy Enterprise Resource Planning (Navy ERP) Program Release 1.1 Acquisition Decision Memorandum (ADM)," *Department of Defense, Office of the Deputy Chief Management Officer,* March 15, 2010.

*McGrath, Elizabeth A., "Memorandum for Director, Report Followup & GAO Liaison, Office of the Inspector General, Department of Defense; Subject: Follow-up on Government Accountability Office Report GAO-08-896, 'DOD Business Systems Modernization: Important Management Controls Being Implemented on Major Navy Program, but Improvements Needed in Key Areas,' September 8, 2008 (GAO Review Code 310659)," *Department of Defense, Office of the Deputy Chief Management Officer,* September 22, 2009.

*McGrath, Elizabeth A., "Memorandum for Office of the Inspector General, Department of Defense (OIG DoD) (Attn: Director, Report Follow-up and GAO Liaison); Subject: Follow-up on Government Accountability Office Report GAO-08-896, 'DOD Business Systems Modernization: Important Management Controls Being Implemented on Major Navy Program, but Improvements Needed in Key Areas,' September 8, 2008 (GAO Review Code 310659)," *Department of Defense, Office of the Deputy Chief Management Officer,* July 30, 2010.

*"Memorandum for the Assistant Secretary of the Navy (Research, Development and Acquisition) (ASN(RDA)); Subject: Navy Enterprise Resource Planning (Navy ERP) Program, Financial and Acquisition Solution, Phased Deployment to Naval Sea Systems Command (NAVSEA) Acquisition Decision Memorandum (ADM)," *Department of Defense, Office of the Deputy Chief Management Officer,* September 24, 2010.

*"Navy Enterprise Resource Planning Critical Change Report," *US Navy, Deputy Assistant Secretary of the Navy Command, Control, Communications and Computers Intelligence, Information Operations, and Space (DASN(C4I/IO/Space)),* August 28, 2009.

*"Navy Enterprise Resource Planning (ERP)," *Office of the Under Secretary of Defense (Acquisition, Technology & Logistics),* Document Acquisition Management Information Retrieval (DAMIR), Major Automated Information System Annual Report (MAIS), Approval Date: December 31, 2007.

*"Navy Enterprise Resource Planning (ERP)," *Office of the Under Secretary of Defense (Acquisition, Technology & Logistics),* Document Acquisition Management Information Retrieval (DAMIR), Major Automated Information System Quarterly Report (MAIS), Approval Date: June 30, 2008.

*"Navy Enterprise Resource Planning (ERP)," *Office of the Under Secretary of Defense (Acquisition, Technology & Logistics),* Document Acquisition Management Information Retrieval (DAMIR), Major Automated Information System Quarterly Report (MAIS), Approval Date: September 30, 2008.

*"Navy Enterprise Resource Planning (ERP)," *Office of the Under Secretary of Defense (Acquisition, Technology & Logistics),* Document Acquisition Management Information Retrieval (DAMIR), Major Automated Information System Annual Report (MAIS), Approval Date: December 31, 2008.

*"Navy Enterprise Resource Planning (ERP)," *Office of the Under Secretary of Defense (Acquisition, Technology & Logistics),* Document Acquisition Management Information Retrieval (DAMIR), Major Automated Information System Quarterly Report (MAIS), Approval Date: December 31, 2008.

*"Navy Enterprise Resource Planning (ERP)," *Office of the Under Secretary of Defense (Acquisition, Technology & Logistics),* Document Acquisition Management Information Retrieval (DAMIR), Major Automated Information System Quarterly Report (MAIS), Approval Date: March 31, 2009.

*"Navy Enterprise Resource Planning (ERP)," *Office of the Under Secretary of Defense (Acquisition, Technology & Logistics),* Document Acquisition Management Information Retrieval (DAMIR), Major Automated Information System Quarterly Report (MAIS), Approval Date: June 30, 2009.

*"Navy Enterprise Resource Planning (ERP)," *Office of the Under Secretary of Defense (Acquisition, Technology & Logistics),* Document Acquisition Management Information Retrieval (DAMIR), Major Automated Information System Quarterly Report (MAIS), Approval Date: September 30, 2009.

*"Navy Enterprise Resource Planning (ERP)," *Office of the Under Secretary of Defense (Acquisition, Technology & Logistics),* Document Acquisition Management Information Retrieval (DAMIR), Major Automated Information System Annual Report (MAIS), Approval Date: December 31, 2009.

*"Navy Enterprise Resource Planning (ERP)," *Office of the Under Secretary of Defense (Acquisition, Technology & Logistics),* Document Acquisition Management Information Retrieval (DAMIR), Major Automated Information System Quarterly Report (MAIS), Approval Date: December 31, 2009.

*"Navy Enterprise Resource Planning (ERP)," *Office of the Under Secretary of Defense (Acquisition, Technology & Logistics),* Document Acquisition Management Information Retrieval (DAMIR), Major Automated Information System Quarterly Report (MAIS), Approval Date: March 31, 2010.

*"Navy Enterprise Resource Planning (ERP)," *Office of the Under Secretary of Defense (Acquisition, Technology & Logistics),* Document Acquisition Management Information Retrieval (DAMIR), Major Automated Information System Quarterly Report (MAIS), Approval Date: June 30, 2010.

*"Navy Enterprise Resource Planning (ERP)," *Office of the Under Secretary of Defense (Acquisition, Technology & Logistics),* Document Acquisition Management Information Retrieval (DAMIR), Major Automated Information System Quarterly Report (MAIS), Approval Date: September 30, 2010.

*"Navy Enterprise Resource Planning (ERP) Program," *US Navy, Navy Enterprise Resource Planning (ERP) Program Office,* September 20, 2010.

*"Navy Enterprise Resource Planning (ERP) Program Overview for John Quirk, Chris Paul, Pablo Carrillo, Vickie Plunkett, and Lynn Williams," *US Navy, Navy Enterprise Resource Planning (ERP) Program Office,* March 25, 2010.

*"Navy Enterprise Resource Planning (Navy ERP) Background Paper," *US Navy, Navy Enterprise Resource Planning (ERP) Program Office,* August 1, 2010.

*"Navy Enterprise Resource Planning (Navy ERP): Hon. Ms. McGrath US House Committee Hearing," *US House of Representatives,* September 29, 2010.

*"Navy ERP," *Office of the Under Secretary of Defense (Acquisition, Technology & Logistics),* Document Acquisition Management Information Retrieval (DAMIR), Acquisition Program Baseline (APB), Approval Date: August 26, 2004.

*"Navy ERP," *Office of the Under Secretary of Defense (Acquisition, Technology & Logistics),* Document Acquisition Management Information Retrieval (DAMIR), Acquisition Program Baseline (APB), Approval Date: December 12, 2006.

*"Navy ERP," *Office of the Under Secretary of Defense (Acquisition, Technology & Logistics),* Document Acquisition Management Information Retrieval (DAMIR), Acquisition Program Baseline (APB), Approval Date: September 19, 2007.

*"Navy ERP," *Office of the Under Secretary of Defense (Acquisition, Technology & Logistics),* Document Acquisition Management Information Retrieval (DAMIR), Proposed Acquisition Program Baseline (APB), Approval Date: March 16, 2010.

*"Program Schedule," *US Navy, Navy Enterprise Resource Planning (ERP) Program Office,* December 2, 2010.

*"Report to Congressional Report to Congressional Requesters: DoD Business Transformation: Improved Management Oversight of Business System Modernization Efforts Needed," *United States Government Accountability Office (GAO),* GAO-11-53, October 2010.

*"Report to Congressional Report to Congressional Requesters: DoD Business Systems Modernization: Navy ERP Adherence to Best Business Practices Critical to Avoid Past Failures," *United States Government Accountability Office (GAO),* GAO-05-858, September 2005.

*"Report to Congressional Report to Congressional Requesters: DoD Business Systems Modernization: Navy Implementing a Number of Key Management Controls on Enterprise Resource Planning System, but Improvements Still Needed," *United States Government Accountability Office (GAO),* GAO-09-841, September 2009.

*"Report to the Subcommittee on Readiness and Management Support, Committee on Armed Services, U.S. Senate: DoD Business Systems Modernization: Important Management Controls Being Implemented on Major Navy Program, but Improvements Needed in Key Areas," *United States Government Accountability Office (GAO),* GAO-08-896, September 2008.

*Stackley, Sean, "Memorandum for Department of Defense Assistant Deputy Chief Management Officer; Subject: Navy Enterprise Resource Planning (Navy ERP) Program Release 1.1, Wholesale and Retail Supply, to Naval Supply Systems Command (NAVSUPSYSCOM) Deployment Recommendation," *Department of Defense, Office of the Assistant Secretary of the Navy (Research, Development and Acquisition),* February 24, 2010.

*"Statement of Mr. Eric Fanning, Deputy Under Secretary of the Navy/Deputy Chief Management Officer Before the United States Senate Committee on Homeland Security and Government Affairs Subcommittee on Federal Financial Management, Government Information, Federal Services, and International Security," *US Senate,* September 29, 2010.

Interviews

Business Transformation Agency (5 November 2010)
Navy Office of the Assistant Secretary (FM&C) (5 November 2010)
Navy ERP – FMO Discussion (9 November 2010)
Navy ERP – NAVAIR (23 November 2010)
Navy ERP – Ron Rosenthal, former PM (1 December 2010)
Navy ERP – Program Management Office (2 December 2010)
Navy ERP – NAVSUP (5 December 2010)
Navy ERP – Tim Thate, DASN(C4I) (8 December 2010)
Navy ERP – Ron Rosenthal, former PM follow-up (10 December 2010)
Navy ERP – SAP, RADM Cowley (14 December 2010)

Data Sources for the Wideband Global Satellite

Note: An asterisk indicates a key source used in the Root Cause Analysis.

*Aldridge, Jr., E. C., "Memorandum for the Secretaries of the Military Departments; Under Secretary of Defense (Comptroller); Assistant Secretary of Defense (C3I); Under Secretary of the Air Force/Director; National Reconnaissance Office: Subject: Delegation of Milestone Decision Authority for DoD Space Systems," *The Under Secretary of Defense (Acquisition, Technology, and Logistics),* February 4, 2002.

Bliss, Gary R., "PARCA Agenda for LockMart 5 April ITP#3 Trip," *PARCA,* April 5, 2010.

Brinton, Turner, "Future of Air Force's WGS Constellation Still Undecided," *Space News International,* January 26, 2009.

Brinton, Turner, "T-Sat Demise Poses Bandwidth Challenge," *Space News,* April 21, 2009.

Chaplain, Cristina T., "Testimony Before the Subcommittee on Strategic Forces, Committee on Armed Services, U.S. Senate: Space Acquisitions: DOD Faces Substantial Challenges in Developing New Space Systems," *United States Government Accountability Office (GAO),* GAO-09-705T, May 20, 2009.

Chaplain, Cristina T., "Briefing on Commercial and Department of Defense Space System Requirements and Acquisition Practices," *United States Government Accountability Office (GAO),* GAO-10-315R, August, 2009.

*"Cost Data Summary Report," *Wideband Global SATCOM Program,* September 6, 2007.

*"Cost Data Summary Report," *Wideband Global SATCOM Program,* December 5, 2008.

*"Cost Data Summary Report," *Wideband Global SATCOM Program,* September 24, 2009.

*"Cost Effectiveness of Wideband Global SATCOM," *Boeing,* June 2007.

*"Cost Performance Report: Boeing Satellite Systems," *Wideband Global SATCOM Program,* January 2010.

*"DCMA Internal Program Status Chart: Wideband Global Satcom," *Defense Contract Management Agency,* date unknown.

*Donley, Michael B., "Letters of Notification to Congress," *US Air Force, Secretary of the Air Force,* March 3, 2010.

*"Exhibit R-2, RDT&E Budget Item Justification: PB 2011 Air Force," *US Air Force,* February 2010.

Forest, Benjamin D., "An Analysis of Military Use of Commercial Satellite Communications," *Naval Postgraduate School,* September 2008.

*Gansler, J. S., "Memorandum for Secretaries of the Military Departments Attn: Acquisition Executives; Subject: Acquisition Decision Memorandum for Wideband Gapfiller Satellite (WGS) Program," *The Under Secretary of Defense (Acquisition and Technology),* December 15, 2000.

*Gianelli, Mike, "Space and Intelligence Systems Presentation to the Government Accountability Office," *Boeing,* December 5, 2005. Not releasable to the general public.

*Hura, Myron, Manuel Cohen, Elliot Axelband, Richard Mason, Mel Eisman, and Shelley H. Wiseman, "Improving USAF Space Acquisition Programs," *RAND Corporation,* unpublished research. Not releasable to the general public.

*Krieg, Kenneth J., "Memorandum for Secretary of the Air Force: Subject: Redelegation of Milestone Decision Authority (MDA)," *The Under Secretary of Defense (Acquisition, Technology, and Logistics),* January 4, 2006.

Lemkin, Bruce S., and Mark J. Reynolds, "Memorandum of Understanding Between the Department of Defense of the United States of America and the Department of Defence of Australia Concerning Joint Production, Operations, and Support of Wideband Global Satellite Communications," *US Air Force, Wideband Global Satellite Program,* 07-284, November 14, 2007.

Magnuson, Stew, and Sandra I. Erwin, "Promise of 'Revolution' in Satellite Communications Faces Challenges," *National Defense,* January 2008.

*McNicol, David L., "Memorandum for Mr. John Landon: Subject: Cost Analysis Improvement Group (CAIG) Report for Wideband Gapfiller Satellite (WGS) Program Milestone II/III Review," *Cost Analysis Improvement Group,* November 4, 2000.

*Melvin, William P., "Wideband Global Satcom: Nunn-McCurdy Executive Summary," *SCMO Director,* April 14, 2010.

*"Notes from the WGS IPT#5 meeting of 2 April at 1330," *The RAND Corporation,* April 2, 2010.

"Notice of Contract Action for WGS B2FO Effort: Solicitation Number: FA8808-10-C-0001," FEDBIZOPPS.GOV, *US Air Force,* January 27, 2010.

*"Nunn-McCurdy Unit Cost Reporting: Kick-Off Meeting," *WGS Nunn-McCurdy Certification Executive Working Group (EWG),* March 26, 2010.

Sirak, Michael C., "Communications Difference," *Air Force Magazine Daily Report,* October 8, 2009.

Spencer, Maj Gen Larry O., "United States Air Force FY 2009 Performance Based Budget Overview," *US Air Force, SAF/FMB,* February 2008.

"State of the Satellite Industry Report," *Futron Corporation,* June 2009.

"The Boeing Company 2009 Annual Report," *Boeing,* 2009.

*"Typical IPT #5 Questions (by Evaluation Area)," *Office of the Under Secretary of Defense (Acquisition, Technology & Logistics),* April 1, 2010.

*Warcholik, Lt Col Ken, "Wideband Global SATCOM (WGS) Status Brief for IPT #5," *US Air Force, Secretary of the Air Force,* April 1, 2010.

*"WGS Nunn-McCurdy Assessment Executive Working Group (EWG): Principals' Meeting," *Office of the Under Secretary of Defense (Acquisition, Technology & Logistics),* April 7, 2010.

"WGS," *Office of the Under Secretary of Defense (Acquisition, Technology & Logistics),* Document Acquisition Management Information Retrieval (DAMIR), Acquisition Program Baseline (APB), Approval Date: December 15, 2000.

"WGS," *Office of the Under Secretary of Defense (Acquisition, Technology & Logistics),* Document Acquisition Management Information Retrieval (DAMIR), Acquisition Program Baseline (APB), Approval Date: February 14, 2003.

"WGS," *Office of the Under Secretary of Defense (Acquisition, Technology & Logistics),* Document Acquisition Management Information Retrieval (DAMIR), Acquisition Program Baseline (APB), Approval Date: February 24, 2004.

"WGS," *Office of the Under Secretary of Defense (Acquisition, Technology & Logistics),* Document Acquisition Management Information Retrieval (DAMIR), Acquisition Program Baseline (APB), Approval Date: April 3, 2007.

"WGS," *Office of the Under Secretary of Defense (Acquisition, Technology & Logistics)*, Document Acquisition Management Information Retrieval (DAMIR), Defense Acquisition Executive Summary (DAES), as of February 1, 2008.

"WGS," *Office of the Under Secretary of Defense (Acquisition, Technology & Logistics)*, Document Acquisition Management Information Retrieval (DAMIR), SAR Baseline, November 21, 2000.

"WGS," *Office of the Under Secretary of Defense (Acquisition, Technology & Logistics)*, Document Acquisition Management Information Retrieval (DAMIR), SAR Baseline, December 16, 2000.

*"WGS," *Office of the Under Secretary of Defense (Acquisition, Technology & Logistics)*, Document Acquisition Management Information Retrieval (DAMIR), Selected Acquisition Report (SAR), September 30, 2001.

*"WGS," *Office of the Under Secretary of Defense (Acquisition, Technology & Logistics)*, Document Acquisition Management Information Retrieval (DAMIR), Selected Acquisition Report (SAR), December 31, 2001.

*"WGS," *Office of the Under Secretary of Defense (Acquisition, Technology & Logistics)*, Document Acquisition Management Information Retrieval (DAMIR), Selected Acquisition Report (SAR), December 31, 2002.

*"WGS," *Office of the Under Secretary of Defense (Acquisition, Technology & Logistics)*, Document Acquisition Management Information Retrieval (DAMIR), Selected Acquisition Report (SAR), June 30, 2003.

*"WGS," *Office of the Under Secretary of Defense (Acquisition, Technology & Logistics)*, Document Acquisition Management Information Retrieval (DAMIR), Selected Acquisition Report (SAR), December 31, 2003.

*"WGS," *Office of the Under Secretary of Defense (Acquisition, Technology & Logistics)*, Document Acquisition Management Information Retrieval (DAMIR), Selected Acquisition Report (SAR), December 31, 2004.

*"WGS," *Office of the Under Secretary of Defense (Acquisition, Technology & Logistics)*, Document Acquisition Management Information Retrieval (DAMIR), Selected Acquisition Report (SAR), September 30, 2005.

*"WGS," *Office of the Under Secretary of Defense (Acquisition, Technology & Logistics)*, Document Acquisition Management Information Retrieval (DAMIR), Selected Acquisition Report (SAR), December 31, 2005.

*"WGS," *Office of the Under Secretary of Defense (Acquisition, Technology & Logistics)*, Document Acquisition Management Information Retrieval (DAMIR), Selected Acquisition Report (SAR), December 31, 2006.

*"WGS," *Office of the Under Secretary of Defense (Acquisition, Technology & Logistics),* Document Acquisition Management Information Retrieval (DAMIR), Selected Acquisition Report (SAR), December 31, 2007.

*"WGS," *Office of the Under Secretary of Defense (Acquisition, Technology & Logistics),* Document Acquisition Management Information Retrieval (DAMIR), Selected Acquisition Report (SAR), December 31, 2009.

*"WGS Nunn-McCurdy Certification: IPT #5 Meeting," *Under Secretary of Defense (Acquisition, Technology & Logistics),* April 1, 2010.

References

Blickstein, Irv, Michael Boito, Jeffrey A. Drezner, James Dryden, Kenneth Horn, James G. Kallimani, Martin C. Libicki, Megan McKernan, Roger C. Molander, Charles Nemfakos, Chad J. R. Ohlandt, Caroline Reilly, Rena Rudavsky, Jerry M. Sollinger, Katharine Watkins Webb, and Carolyn Wong, *Root Cause Analyses of Nunn-McCurdy Breaches,* Volume 1: Zumwalt-*Class Destroyer, Joint Strike Fighter, Longbow Apache, and Wideband Global Satellite,* Santa Monica, Calif.: RAND Corporation, MG-1171/1-OSD, 2011. As of July 1, 2012:
http://www.rand.org/pubs/monographs/MG1171z1.html

Blickstein, Irv, Jeffrey A. Drezner, Martin, C. Libicki, Brian McInnis, Megan McKernan, Charles Nemfakos, Jerry M. Sollinger, and Carolyn Wong, *Root Cause Analyses of Nunn-McCurdy Breaches,* Volume 2: *Excalibur Artillery Projectile and the Navy Enterprise Resource Planning Program, with an Approach to Analyzing Complexity and Risk,* Santa Monica, Calif.: RAND Corporation, MG-1171/2-OSD, 2012. As of August 17, 2012:
http://www.rand.org/pubs/monographs/MG1171z2.html

DoD 5000.4-M, "Cost Analysis Guidance and Procedures," December 11, 1992.

Glass, Robert, *Software Runaways: Monumental Software Disasters,* Upper Saddle River, N.J.: Prentice Hall, 1997.

Goldstein, Harry, "Who Killed the Virtual Case File?" *IEEE Spectrum,* Vol. 42, No. 9, September 2005, pp. 24–35.

Government Accountability Office, *Air Traffic Control: Immature Software Acquisition Processes Increase FAA System Acquisition Risks,* GAO/AIMD-97-47, Washington, D.C., March 1997.

Government Printing Office, "Congressional Hearings," undated.

Ike Skelton Defense Authorization Act of Fiscal Year 2011, December 20, 2010.

McCaney, Kevin, "Readers Offer Stuffing for IT Turkey," *Federal Computer Week,* December 1, 2009. As of February 9, 2010:
http://fcw.com/Articles/2009/12/01/IT-turkeys-redux-readers-comments.aspx?

Naval Center for Cost Analysis, "NCCA Inflation Indices," January 2010. As of May 14, 2011:
http://www.ncca.navy.mil/services/inflation.cfm

Public Law 111-23, Weapon Systems Acquisition Reform Act (WSARA) of 2009, May 22, 2009.